ELECTRICITY GENERATION USING WIND POWER

ELECTRICITY GENERATION USING WIND POWER

William Shepherd
University of Bradford, UK

Li Zhang
University of Leeds, UK

NEW JERSEY · LONDON · SINGAPORE · BEIJING · SHANGHAI · HONG KONG · TAIPEI · CHENNAI

Published by

World Scientific Publishing Co. Pte. Ltd.
5 Toh Tuck Link, Singapore 596224
USA office: 27 Warren Street, Suite 401-402, Hackensack, NJ 07601
UK office: 57 Shelton Street, Covent Garden, London WC2H 9HE

British Library Cataloguing-in-Publication Data
A catalogue record for this book is available from the British Library.

ELECTRICITY GENERATION USING WIND POWER

Copyright © 2011 by World Scientific Publishing Co. Pte. Ltd.

All rights reserved. This book, or parts thereof, may not be reproduced in any form or by any means, electronic or mechanical, including photocopying, recording or any information storage and retrieval system now known or to be invented, without written permission from the Publisher.

For photocopying of material in this volume, please pay a copying fee through the Copyright Clearance Center, Inc., 222 Rosewood Drive, Danvers, MA 01923, USA. In this case permission to photocopy is not required from the publisher.

Desk Editor: Tjan Kwang Wei

ISBN-13 978-981-4304-13-9
ISBN-10 981-4304-13-1

Typeset by Stallion Press
Email: enquiries@stallionpress.com

Printed by FuIsland Offset Printing (S) Pte Ltd. Singapore

Foreword and Acknowledgement

This book is written for electrical engineers and students of electrical engineering. As a textbook it is pitched at the level of final-year undergraduates and postgraduates. There is no detailed coverage of the aeronautical and meteorological features of wind turbines. The book is not intended as a design handbook. Certain of the chapters contain end-of-chapter numerical problems, with the answers shown separately at the end of the book. Some of the material in chapters 2, 5, 6 and 7 is reworked from earlier publications by of one of the authors (WS). This material is acknowledged in appropriate places and the authors are grateful to the publishers of the earlier work for their permission to reproduce it.

Bradford, England
2010

Contents

	Foreword and Acknowledgement		v
1.	The Development of Wind Converters		1
	1.1	Nature and Origin of the Wind	1
	1.2	Development of Wind Converters	3
	References		6
2.	Theory of Wind Converters		7
	2.1	Power and Energy Basis of Wind Converters	7
		2.1.1 Origin and properties of the wind	7
		2.1.2 Power and energy	8
	2.2	Theoretical Power Available in the Wind	9
	2.3	Theoretical Maximum Power Extractable from the Wind	11
	2.4	Practical Power Extractable from the Wind	15
		2.4.1 Power coefficient	15
		2.4.2 Torque versus rotational speed	16
		2.4.3 Shaft power versus rotational speed	16
		2.4.4 Tip-speed ratio (TSR)	17
	2.5	Mechanical Features of Wind Machines	19
		2.5.1 Axial thrust (Pressure)	19
		2.5.2 The "Yaw" effect	20
		2.5.3 Gyroscopic forces and vibrations	20
		2.5.4 Centrifugal forces	22
		2.5.5 Solidity factor	22

		2.5.6	Two rotor blades or three rotor blades?	23
		2.5.7	Shaft torque and power	24
	2.6	Fixed Rotational Speed or Variable Rotational Speed?		26
		2.6.1	Constant speed operation	27
		2.6.2	Variable speed operation	28
	2.7	Efficiency Considerations of Wind-Powered Electricity Generation		29
	2.8	Worked Numerical Examples on Wind-Turbine Operation		31
	2.9	Problems and Review Questions		36
	References			38
3.	Past and Present Wind-Energy Turbines			41
	3.1	Nineteenth-Century Windmills		41
	3.2	Early Twentieth-Century Wind-Energy Turbines		43
	3.3	Later Twentieth-Century Wind-Energy Turbines		48
	3.4	Modern Large Wind Power Installations		51
	3.5	Worked Numerical Example		59
	3.6	Vertical Axis Wind Machines		60
		3.6.1	The Savonius design	61
		3.6.2	The Darrieus design	62
		3.6.3	Other forms of vertical axis machine	63
	References			63
4.	The Location and Siting of Wind Turbines			65
	4.1	The Availability of Wind Supply		65
		4.1.1	Global survey	65
		4.1.2	Energy content of the wind	66
		4.1.3	Wind-energy supply in Europe	68
		4.1.4	Wind-energy supply in the USA	74
	4.2	Statistical Representation of Wind Speed		79
	4.3	Choice of Wind Turbine Sites		84
		4.3.1	Identification of suitable areas	85
		4.3.2	Selection of possible sites within the chosen area	85

	4.4	Effects of the Site Terrain	87
	4.5	Spacing Effects of Wind Farm Arrays	89
	4.6	Problems and Review Questions	91
	References	92
5.	Power Flow in Electrical Transmission and Distribution Systems	93
	5.1	Basic Forms of Power Transmission Networks	93
	5.2	Current and Voltage Relationships	95
	5.3	Power Relationships in Sinusoidal Circuits	99
		5.3.1 Instantaneous power	99
		5.3.2 Average power and apparent power	100
		5.3.3 Power factor	101
		5.3.4 Reactive power	103
	5.4	Complex Power	105
	5.5	Real Power Flow and Reactive Power Flow in Electrical Power Systems	109
		5.5.1 General summary	109
		5.5.2 Summary from the perspective of the consumer	111
	References	111
6.	Electrical Generator Machines in Wind-Energy Systems	113
	6.1	DC Generators	113
	6.2	AC Generators	114
	6.3	Synchronous Machine Generators	114
	6.4	Three-Phase Induction Machine	121
		6.4.1 Three-phase induction motor	122
		6.4.2 Three-phase induction generator	127
		6.4.3 Different generation systems	132
	6.5	Analysis of Induction Generator in Terms of Complex Vector Representation	136
		6.5.1 Three-phase to d-q-0 space vector transformation	140
	6.6	Switched Reluctance Machines	143
		6.6.1 Switched reluctance motors	143
		6.6.2 Switched reluctance generator	144

	6.7	What Form of Generator is the Best Choice for Wind Generation Systems?	145
	References		146

7. Power Electronic Converters in Wind-Energy Systems   147

- 7.1 Types of Semiconductor Switching Converters   147
- 7.2 Three-Phase Controlled Bridge Rectifier   148
- 7.3 Three-Phase Controlled Bridge Inverter Feeding an Infinite Bus   154
  - 7.3.1 Output voltage   154
  - 7.3.2 Real (average) power output   158
  - 7.3.3 Reactive power   159
  - 7.3.4 RMS output current   160
  - 7.3.5 Inverter power factor   162
- 7.4 The Effect of AC System Reactance on Inverter Operation   164
- 7.5 Three-Phase Cycloconverter Feeding an Infinite Bus   165
- 7.6 Matrix Converter Feeding an Infinite Bus   166
- 7.7 Worked Numerical Examples   169
  - 7.7.1 Three-phase bridge rectifier   169
  - 7.7.2 Three-phase bridge inverter feeding on infinite bus   170
- 7.8 Commonly Used Forms of Power Electronic Drive in Wind-Energy Systems   175
  - 7.8.1 Fixed-speed and directly coupled cage induction generator   175
  - 7.8.2 Variable-speed and doubly fed induction generator   176
  - 7.8.3 Variable-speed and direct drive synchronous generator   177
- 7.9 Problems and Review Questions   178
  - 7.9.1 Three-phase controlled bridge rectifier, with ideal supply, feeding a highly inductive load . .   178

	7.9.2	Three-phase, full-wave, and controlled bridge inverter feeding an infinite bus	179
	References		180

8. Integrating Wind Power Generation into an Electrical Power System — 181

- 8.1 Electricity Distribution Systems 182
- 8.2 Issues for Consideration Concerning the Integration of Wind-Energy Generation into an Electric Power System 183
  - 8.2.1 Energy credit 184
  - 8.2.2 Capacity credit 187
  - 8.2.3 Control and reliability 188
- 8.3 The Effect of Integrated Wind Generation on Steady-State System Voltages 190
- 8.4 The Effect of Integrated Wind Generation on Dynamic and Transient System Voltages 193
  - 8.4.1 Lightning strikes 194
  - 8.4.2 Voltage flicker 194
  - 8.4.3 Harmonics 195
  - 8.4.4 Self-excitation of induction generators 200
- References 201

9. Environmental Aspects of Wind Energy — 203

- 9.1 Reduction of Emissions 203
  - 9.1.1 World consumption of coal 203
  - 9.1.2 Open coal fires 205
- 9.2 Effluents due to Coal Burning 206
  - 9.2.1 Sulphur oxides 206
  - 9.2.2 Nitrogen oxides 207
  - 9.2.3 Particulates 208
  - 9.2.4 Carbon dioxide 209
- 9.3 Wind Turbine Noise 209
  - 9.3.1 Measurement of wind turbine aerodynamic noise 212
  - 9.3.2 Mechanical noise 214

xii *Electricity Generation Using Wind Power*

	9.4	Electromagnetic Interference from Wind Turbines	215
		9.4.1 Electromagnetic interference radiated from wind turbines	215
		9.4.2 Electromagnetic interference effects due to the rotating blades	216
	9.5	Effect of a Wind Turbine on Wildlife	217
	9.6	Visual Impact of Wind Turbines	219
		9.6.1 Individual response	219
		9.6.2 Shadow flicker	219
	9.7	Safety Aspects of Wind-Turbine Operation	220
	References		220
10.	**Economic Aspects of Wind Power**		**223**
	10.1	Investment Aspects of Wind-Powered Electricity Generation	223
		10.1.1 Costs of the turbines and generators	224
		10.1.2 Costs of the turbine site, construction, and grid connection	225
		10.1.3 Operation and maintenance (O and M) costs	226
		10.1.4 Turbine lifetime and depreciation rate	227
		10.1.5 Cost associated with the financing of wind farm building and operation	228
		10.1.6 Wind regime at the turbine site	229
	10.2	Comparative Costs of Generating Electricity from Different Fuel Sources	230
	References		234

Answers to the End of Chapter Problems — 237

Index — 241

CHAPTER 1

The Development of Wind Converters

1.1 Nature and Origin of the Wind

The wind is the motion of a mass of air. For the purpose of using wind energy, it is normally the horizontal component of the wind that is of interest. There is also a vertical component of the wind that is very small compared with the horizontal component, except in local disturbances such as thunderstorm updrafts.

At the earth surface, the atmospheric pressure is measured in the unit Pascal (Pa) and has an average value 101,325 Pa, which is sometimes called "one atmosphere". Another unit of pressure used for meteorological calculations is the millibar (mbar). There are exactly 100 Pa per millibar so that one atmosphere is about 1,000 mbar. On a map, regions of equal atmospheric pressure are identified by isobar lines such as those illustrated in Fig. 1.1. A close concentration of isobar lines indicates a high pressure gradient or region of rapid pressure change. Wind speed is directly proportional to the pressure gradient.

The atmospheric pressure varies from place-to-place and from day-to-day, caused by the combined effects of solar heating and the rotation of the earth. As the earth spins, illustrated in Fig. 1.2, the atmospheric air

2 *Electricity Generation Using Wind Power*

Fig. 1.1   Atmospheric pressure isobars for North America, April 2008[1].

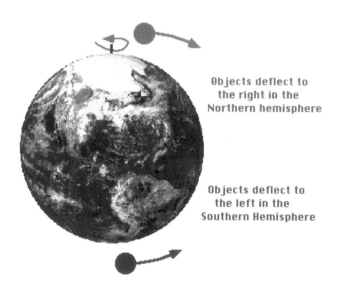

Fig. 1.2   The Coriolis effect on wind direction[1].

surrounding it is dragged round with it at different levels depending on altitude. The mix of air forms turbulence causing wind at the earth surface.

An additional feature is the inertial force known as the Coriolis force, which occurs in rotational systems. When air moves over the surface of the earth as it rotates, instead of travelling in a straight line, the path of the moving air veers to the right. The effect is that air moving from an area of higher pressure to an area of lower pressure moves almost parallel to the isobars. In the northern hemisphere, the wind circles in a clockwise direction towards the area of low pressure but in the southern hemisphere, the wind circles in an anti-clockwise direction, as shown in Fig. 1.2.

The heating effect of solar radiation varies with latitude and with the time of day. The warming effect is greater over the equator causing less dense warmer air to rise above the cooler air, reducing the surface atmospheric pressure compared with the polar regions. The combined effect of the solar heating and the Coriolis force is to create the following prevailing wind directions[1].

Latitude	90°–60° N	60°–30° N	30°–0° N	northern
Direction	NE	SW	NE	hemisphere
Latitude	0°–30°	30°–60°	60°–90°	southern
Direction	SE	NW	SE	hemisphere

In any particular location, the wind direction is much influenced by the presence of land masses and features such as mountain ranges. The general picture of sea winds for the north and south Atlantic regions is shown in Fig. 1.3[1].

1.2 Development of Wind Converters

Wind energy provided the motive power for sailing ships for thousands of years, until the age of steam. The fortunes of the European colonial powers such as England, France, Germany, Spain, Portugal, Holland and Belgium rested on their mastery of the sea and its navigation. But the intermittent nature and uncertain availability of the wind combined with the relative slowness of wind powered vessels gradually gave way to fossil-fuel powered commercial shipping. Today, most shipping uses oil fuelled diesel engines.

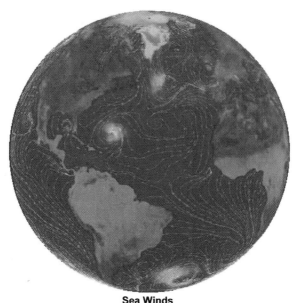

Sea Winds
Credit: NASA JPL Satellite: QuikSCAT Sensor: SeaWinds

Fig. 1.3   Global wind distribution[1].

However, yachting and small boat sailing remain important recreational sports throughout the world.

The wind has also been used for thousands of years to provide the motive power for machines acting as water pumps or used to mill or grind grain. Such machines came to be known as "windmills". The operators of windmills in feudal England took the name of their craft and acquired the surname Miller (or Millar).

Very early wind machines were vertical axis structures and have been identified in China, India, Afghanistan and the Middle East, especially Persia, going back to about 250 B.C. and possibly much earlier.

Horizontal axis wind machines were developed by the Arab nations and their use became widespread throughout the Islamic World. In Europe, the horizontal axis wind machine became established about the 11th Century A.D., mostly having the form of a tower and rotating sails which became known as the Dutch windmill. The earliest recorded windmill dates from 1191 A.D. By the 18th Century A.D., multi-sail Dutch windmills were extensively used in Europe. It is estimated that by 1750 A.D. there were

Fig. 1.4 Classical 'Dutch' windmill in Southern England (unknown origin).

8,000 windmills in operation in Holland and about 10,000 windmills in both Britain and Germany. Dutch settlers built windmills in North America, mainly along the eastern coast areas that became the New England states of the USA. At one stage, the shore of Manhattan Island was lined with windmills built by Dutch settlers[2].

The principal features of the classical type of Dutch windmill are illustrated in Fig. 1.4. Usually there are four sails, located upstream (i.e., facing into the wind). Five, six and eight mills have been built. Although a five-sailed machine is relatively efficient, it is disabled by the failure of just one of its blades. On the other hand, a six-sailed machine can continue to operate with four, three or two sails, if necessary. One of the many engineering sketches left by Leonardo da Vinci (1452–1519 A.D.) represents a design for a six-sailed windmill[3].

To achieve variation of the rotational speed, the effective sail area of a Dutch type of windmill can be modified by the use of shutters. This corresponds to furling the sails on a yacht. Furling or shuttering can also be used to prevent over-speeding in high wind conditions. The cupola on top of the tower, in Fig. 1.4 for example, is designed to rotate, under the guidance of a rudder or stabilizer wheel, so that the sails remain upstream and perpendicular to the wind direction. Mechanical rotational power obtained from the sail shaft is transmitted down the tower, via a beveled toothed bearing, onto a vertical drive shaft. This, in turn, drives a toothed gear system which supplies power to rotate a grinding wheel for the corn.

In Fig. 1.4, the height of the horizontal, rotating axis above the ground is often called the "hub height". For a typical windmill this might be 30 ft, 40 ft or even 50 ft high. Despite such a large structure the power rating of this Dutch windmill is the mechanical equivalent of only a few tens of kilowatts. The power developed by such a large structure is therefore roughly equivalent to the electrical power supply now required by a large family house in Western Europe or North America. In engineering terms, the efficiency of a Dutch windmill is low, although this may be a secondary consideration since the input power is free.

Wind energy is transmitted by what is essentially a low density fluid. (i.e., the wind). The physical dimensions of any device used to convert its kinetic energy into a usable form are necessarily large in relation to the power produced. Wind availability is not only intermittent but unpredictable. The energy source, however, is free, environmentally clean and infinitely renewable. There is no pollution and no direct use of fossil fuels in the energy gathering process.

References

1. "Wind Energy and Wind Power", Solcomhouse website, Oct. 2008, http://www.solcomhouse.com/windpower.htm
2. McVeigh, J. C., *Energy Around the World*, Pergamon Press, Oxford, England, 1984.
3. Golding, E. W. *The Generation of Electricity by Wind Power* Chapter 2 "The History of Windmills", E. and F. N. Spon Ltd., London, England, 1955.

CHAPTER 2
Theory of Wind Converters

2.1 Power and Energy Basis of Wind Converters

2.1.1 Origin and properties of the wind

From the viewpoint of energy conversion, the most important properties of the wind at a particular location are the velocity of the airstream and the air density. The air density varies with altitude and with atmospheric conditions such as temperature, pressure, and humidity. At sea level and at standard atmospheric temperature and pressure, the value is:

$\rho = 1.201 \text{ kg/m}^3$ at 1,000 millibars (29.53 inches of mercury) or 101.3 kilo pascal (kPa) pressure and temperature 293 K.

In the UK, a useful figure for the atmospheric air density is:

$$\rho = 1.29 \text{ kg/m}^3 (0.08 \text{ lb/ft}^3). \qquad (2.1)$$

In the USA, a commonly quoted figure, for sea level under dry conditions at a temperature of 0°C (273 K), is:

$$\rho = 1.275 \text{ kg/m}^3. \qquad (2.2)$$

There is a lot of local variation of the values of the air density in different areas of the world. It is found that the temperature, pressure, and density of the air decrease with altitude. For wind-turbine applications, the range of interest is mostly within a couple of hundred feet of ground level. Within this range, it is adequate to use the density values in Eqs. (2.1) and (2.2) or their local alternatives at other locations.

2.1.2 *Power and energy*

It is important to note that power P and energy W are not the same thing. The energy of a system is its capacity for doing work, irrespective of the time taken to do it. The power of a system is the time rate of doing work or expending energy and therefore has the dimension of energy (or work) divided by time.

$$\text{Power} = \frac{\text{Energy (Work)}}{\text{Time}}. \qquad (2.3)$$

For large increments of time t, the average power P is given by:

$$P = \frac{W}{t}. \qquad (2.4)$$

For small increments of time dt, the instantaneous value of the power P_{inst} is given by:

$$P_{\text{inst}} = \frac{dW}{dt}. \qquad (2.5)$$

In most wind-energy calculations, the average power in Eq. (2.4) is used.

In the Systeme Internationale (S.I.) system of units used in this book, the unit of energy is the joule and the unit of power is the joule per sec (J/s), which is usually called the watt (W). For practical engineering purposes, it is often more convenient to use the kilowatt (kW) or megawatt (MW). A table of conversion factors relating power and energy is given as Table 2.1.

Power in watts is not concerned exclusively with electrical engineering. For example, the rotational mechanical power of engines is often expressed in kilowatts. The power ratings in watts of various devices and animals are shown in Fig. 2.1, using a logarithmic scale.[1]

In terms of the human perception of power, it is sometimes helpful to use the old British power unit of horsepower (HP).

$$1 \text{ Horsepower (HP)} = 746 \text{ W}. \qquad (2.6)$$

Energy converters with a rotational mechanical output, such as combustion engines and wind turbines, can be rated either in the mechanical units of

Table 2.1 Conversion factors in power and energy.

	Unit	Equivalents
Power	1 watt (W)	1 joule/sec (j/s) = 0.001341 HP
	1 kilowatt (kW)	1000 W = 1.34 HP
	1 megawatt (MW)	1,000,000 W = 1340 HP
	1 horsepower	745.7 W = 550 ft lb/sec
Power density	1 w/m^2	3.6 kJ/m^2/h
Energy	1 Joule	1 watt-second (Ws)
	1 kilowatt hour (kWh)	3.6×10^6 Joules
	1 megawatt hour (MWh)	3.6×10^9 Joules
Pressure	1 Pascal (Pa)	1 N/m^2
	1 bar	10^5 Pa = 14.5 lb/m^2
	1 lb/in^2 (psi)	6.895 kPa
	Atmospheric Pressure = 14.7 psi	101.325 kPa

horsepower or in the electrical units of kilowatts. From Fig. 2.1, for example, an automobile is shown as having an efficiency of about 25% and has a power rating of about 0.3×10^6 W, which is roughly equivalent to 400 HP.[1]

2.2 Theoretical Power Available in the Wind[1]

If the air mass is m and it moves smoothly with an average velocity V, the motion of the air mass has a kinetic energy (KE).

$$\text{KE} = \frac{1}{2}mV^2. \quad (2.7)$$

Consider a smooth and laminar flow of wind passing perpendicularly (normally) through an element of area A of any shape, having thickness x, shown in Fig. 2.2.

The mass m of air contained in an element of volume xA is given, in terms of density ρ, by:

$$m = \rho A x. \quad (2.8)$$

Combining Eqs. (2.7) and (2.8) gives, for the KE associated with this mass and volume of air:

$$\text{KE} = \frac{1}{2}\rho A x V^2. \quad (2.9)$$

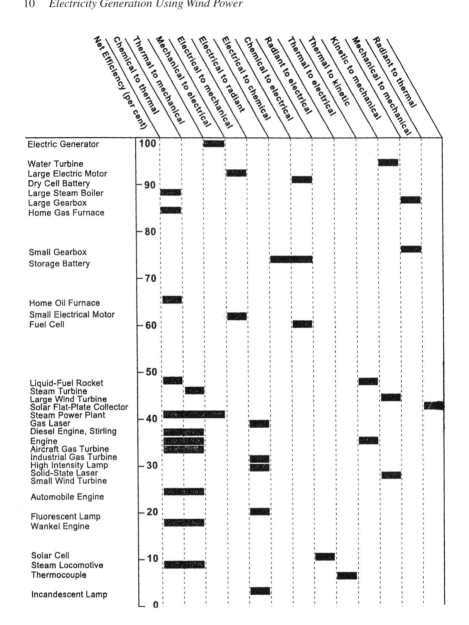

Fig. 2.1   Efficiencies of energy converters.[1]

Table 2.2 Power available in the wind³ (at standard temperature and pressure).

Power	Swept area, A	Wind velocity, V	Power equation, P_w
W	ft²	mph	$0.0053\,AV^3$
W	m²	m/s	$0.64\,AV^3$

It is seen that RHS of Eq. (2.9) represents a force $(1/2)\rho A V^2$ multiplied by a distance x. Now, the KE passing through the element per unit time is equal to the power rating:

$$P_{element} = \frac{d(KE)}{dt} = \frac{1}{2}\rho A x V^2 \frac{dx}{dt}. \qquad (2.10)$$

But the average time rate of change of the displacement, dx/dt, is the average wind velocity V:

$$\frac{dx}{dt} = V. \qquad (2.11)$$

The average power in the wind P_w is obtained by combining Eqs. (2.10) and (2.11) to give:

$$P_W = \frac{1}{2}\rho A V^3. \qquad (2.12)$$

Equation (2.12) is the basis of all wind power and energy calculations. The most obvious feature is that the wind power is proportional to the cube of the average wind speed. It is clear that the average wind speed is, by far, the dominant consideration in wind turbine location.

Table 2.2 summarises more detailed expressions for the power in watts in English units and in metric units.

2.3 Theoretical Maximum Power Extractable from the Wind[1,2]

Only a fraction of the total theoretical power available in the wind, represented by Eq. (2.12), is extractable. It is an intrinsic property of all physical systems that when energy is converted from one form to another, this conversion is accompanied by various energy losses. The result is that conversion is always subject to significant intrinsic limitations of efficiency.

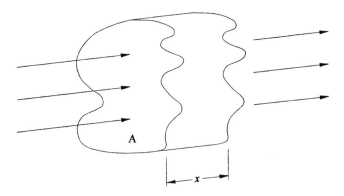

Fig. 2.2 Element of space through which the air flow passes.[1]

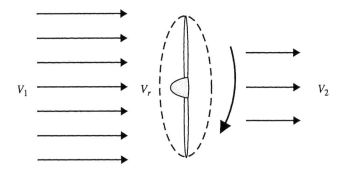

Fig. 2.3 Rotor of a wind converter.[1]

Let a flow of smooth and steady air with an upstream average velocity V_1 impinge upon the rotor of a wind machine, as illustrated in Fig. 2.3. Some of the energy from the wind is transferred to the wind machine rotor so that the smooth and steady air far downstream flows at a smaller average velocity V_2. The KE reduction of the airflow, of mass m, per unit time is:

$$\text{Kinetic Energy (KE)} = \frac{1}{2} m V_1^2 - \frac{1}{2} m V_2^2$$

$$= \frac{1}{2} m \left(V_1^2 - V_2^2 \right). \qquad (2.13)$$

In the process of extracting energy from the wind, the wind velocity V_r that actuates the rotor is less than the upstream "free wind" velocity V_1. With an ideal and lossless system, all of the energy reduction in the airstream is transferred to the rotor of the wind machine. The downstream average velocity V_2 is then smaller than the actuating velocity V_r at the rotor.

Combining Eq. (2.8) with Eq. (2.11) for the airstream at the rotor blades gives an expression for the time rate of air mass transferred.

$$\frac{dm}{dt} = \rho A \frac{dx}{dt} = \rho A V_r. \tag{2.14}$$

The power at the rotor is, from Eq. (2.5), the time rate of KE transferred,

$$P_r = \frac{d(KE)}{dt}. \tag{2.15}$$

Substituting Eqs. (2.13) and (2.14) separately into Eq. (2.15) gives:

$$P_r = \frac{1}{2}\frac{dm}{dt}(V_1^2 - V_2^2) = \frac{1}{2}\rho A V_r (V_1^2 - V_2^2). \tag{2.16}$$

Now the air mass passing through the rotor undergoes not only an energy reduction but a reduction of linear momentum:

$$\text{Reduction of Linear Momentum} = m(V_1 - V_2). \tag{2.17}$$

The time rate of the change of momentum reduction is a force, of value:

$$\frac{d}{dt}(m(V_1 - V_2)) = \frac{dm}{dt}(V_1 - V_2) = \rho A V_r (V_1 - V_2). \tag{2.18}$$

By equating the rate of the change of KE transfer (2.16) with the power associated with the rate of the change of momentum, from Eq. (2.18), it is found that the wind velocity V_r at the rotor may be expressed as:

$$V_r = \frac{V_1 + V_2}{2}. \tag{2.19}$$

In this idealised model of airflow, the wind velocity at the rotor is therefore the average of the upstream and downstream steady wind velocities. Substituting for V_r from Eq. (2.19) into Eq. (2.16) gives an expression for the power extractable from the wind by the rotor. Using the symbol P_{ex} to denote extracted power,

$$P_{ex} = \frac{1}{4}\rho A(V_1 + V_2)(V_1^2 - V_2^2) = \frac{1}{4}\rho A V_1^3 \left[1 + \frac{V_2}{V_1} - \left(\frac{V_2}{V_1}\right)^2 - \left(\frac{V_2}{V_1}\right)^3\right]$$

$$= \frac{1}{4}\rho A V_1^3 \left[1 + \frac{V_2}{V_1}\right]\left[1 - \left(\frac{V_2}{V_1}\right)^2\right]. \tag{2.20}$$

The value of wind velocity ratio V_2/V_1 that results in maximum power transfer is calculated by differentiating Eq. (2.20) with respect to (V_2/V_1)

and equating to zero. This results in a quadratic equation showing that to maximise P_{ex} the ratio V_2/V_1 must have the values either $V_2/V_1 = 1/3$ or $V_2/V_1 = -1$.

The negative option is meaningless so that the correct solution is:

$$\frac{V_2}{V_1} = \frac{1}{3}. \qquad (2.21)$$

Substituting Eq. (2.21) into Eq. (2.20) gives an expression for the maximum possible power extraction, under ideal conditions.

$$P_{ex}(\max) = \frac{8}{27}\rho A V_1^3 = \left(\frac{16}{27}\right)\frac{1}{2}\rho A V_1^3$$

$$= (0.593)\frac{1}{2}\rho A V_1^3. \qquad (2.22)$$

The very important result (2.22) is sometimes referred to as Betz Law, being named after the German astroscientist Albert Betz of Göttingen. This states that "even with ideal wind energy conversion the maximum power transferable is only 0.593 or 16/27 of the total power in the wind". In reality, only a fraction of this theoretical maximum power is realised, even by the best-designed turbines.

Table 2.3 shows the values of the maximum theoretical power obtainable from a range of wind-turbine sizes at typical wind speeds.[3] It is notable how large a circular area must be used to generate any useful amount of power. For example, in a 10-mph wind, which is a light breeze, a swept area of 25 ft diameter would realise only a maximum theoretical value of 1.5 kW (and a practical value of roughly one-half of that). This immediately points

Table 2.3 Maximum theoretical power extractable by ideal wind machine[3] $\left(\frac{0.593}{2}\rho AV^2\right)$.

Wind speed (mph)	Power (kW) from circular area of different diameter (ft)				
	12.5	25.0	50.0	100	200
10	0.38	1.5	6.0	24	96
20	3.08	12.3	49.2	196	784
30	10.4	41.6	166.4	666	2,664
40	24.6	98.4	393.6	1,574	6,296
50	48.2	192.8	771.2	3,085	12,340
60	83.2	332.8	1,331.0	5,326	21,300

to the difficulty of using wind energy for domestic use in urban areas—the swept area required is too large to be practicable.

Wind speed is usually expressed in miles per hour (mph) but also often in metric units of metre/second (m/s). It is useful to use the conversion factor:

$$1 \text{ mph} = 0.447 \text{ m/s}. \tag{2.23}$$

2.4 Practical Power Extractable from the Wind

2.4.1 *Power coefficient*

The power actually extractable by a wind turbine is much less than the maximum theoretical value defined in Eq. (2.22) and illustrated in Table 2.3. A practical wind machine experiences air drag and air friction on the rotor blades causing heat losses. In addition, the rotation of the blades causes swirling of the air and eddies, which reduce the torque imparted to the blades. The net effect of the various losses is incorporated into a parameter called the power coefficient C_p. With an upstream velocity V_1, the extractable power P_{ex} can be written as:

$$P_{ex} = C_p \frac{1}{2} \rho A V_1^3. \tag{2.24}$$

Coefficient C_p is seen to be the ratio of the power P_w in the wind (2.12) and the power P_{ex} extracted (2.24). It therefore represents the efficiency of the turbine rotor:

$$C_p = \frac{P_{ex}}{P_w} = \text{turbine rotor efficiency}. \tag{2.25}$$

Parameter C_p is a dimensionless variable. By comparison of Eq. (2.24) with Eq. (2.20), it is seen that C_p can be expressed in terms of the upstream and downstream average wind speeds.

$$C_p = \frac{1}{2} \left[1 + \frac{V_2}{V_1} \right] \left[1 - \left(\frac{V_2}{V_1} \right)^2 \right]. \tag{2.26}$$

For the ideal theoretical case, when $V_2/V_1 = 1/3$, coefficient C_p has a maximum or Betz Law value of 0.593. But for practical wind turbines, the value is usually in the range $0 \leq C_p \leq 0.4$. With a value $C_p = 0.4$, for example, the power available from the wind is 0.4/0.593 or about 67% of the ideal theoretical value and is 40% of the total power in the wind. Power coefficient C_p has a value that depends on the wind average velocity, the turbine rotational velocity and also on turbine blade design parameters such as the pitch angle.

2.4.2 Torque versus rotational speed

The basic operating characteristic for any rotational mechanical machine is the shaft torque T versus the shaft rotational speed ω. A separate T-ω characteristic is obtained for each different value of wind speed at the turbine rotor. Typical forms of characteristic are shown in Fig. 2.4 for two wind speeds V' and V'', where $V'' > V'$. The value of the torque at zero speed is the starting or stall torque, caused by friction, which has to be overcome before rotation will commence. Stall torque increases, for any wind turbine, as the wind velocity V increases. The locus of the maximum torque values T_m in Fig. 2.4 is a quadratic of the form $T_m = K_1 \omega^2$, where K_1 is a constant.

2.4.3 Shaft power versus rotational speed

Variation of the shaft power P versus rotational speed ω is shown in Fig. 2.5, for two wind speeds V' and V'', where $V'' > V'$. Now, the shaft power is the product of shaft torque T and the shaft speed ω. In S.I. units, there is no

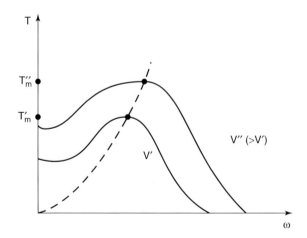

Fig. 2.4 Shaft torque v shaft speed at wind speeds V' and V''.

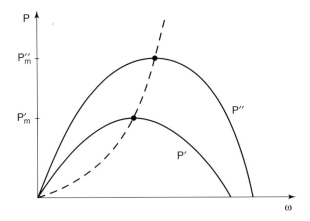

Fig. 2.5   Shaft power v shaft speed at wind and speeds V' and V''.

constant of proportionality so that:

$$P = T\omega. \qquad (2.27)$$

At zero speed, the shaft power is also zero, which is seen to be true for the two cases shown in Fig. 2.5.[4] The shaft power also becomes zero when the shaft torque is zero, illustrated in the compatible diagrams of Figs. 2.4 and 2.5. Maximum shaft power P_m increases with high values of wind speed. The locus of the maximum power points, shown as a dotted line in Fig. 2.5, is a cubic with the form $P_m = K_2\omega^3$, where K_2 is constant. The rotational speed at which maximum power is developed is not, in general, the same as that for which maximum shaft torque occurs.

2.4.4 *Tip-speed ratio (TSR)*

In order to express the power coefficient C_p in terms of both the upstream wind velocity V and the blade rotational velocity ω, a parameter called the tip-speed ratio (TSR) is defined. Figure 2.6 illustrates the main physical parameters, including the blade radius r. The instantaneous velocity v of the blade tip is related to the angular velocity of rotation ω by the relationship:

$$v = r\omega. \qquad (2.28)$$

For a blade of radius r, the TSR is defined as:

$$\text{TSR} = \frac{v}{V} = \frac{r\omega}{V}. \qquad (2.29)$$

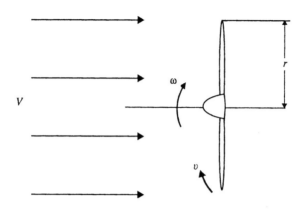

Fig. 2.6 Motion of a two-blade propeller.[1]

The blade rotational velocity n in *rpm* is related to the angular velocity ω in radians per second by the relationship:

$$\omega = \frac{2\pi n}{60}. \tag{2.30}$$

If C_p is plotted against V, there will be a different characteristic for every value of ω. Similarly, if C_p is plotted against ω, there will be a different characteristic for every value of V. The characteristic of C_p versus TSR is a "universal" curve that subsumes values of both ω and V. Good rotor design requires that the maximum value of the power coefficient C_{pm} occurs near to the design-rated value of the rotational speed. Typical characteristics for various different types of wind turbine are shown in Fig. 2.7. The maximum ideal efficiency characteristic for propeller machines is asymptotic to the Betz Law value of 0.593. It can be seen that the most efficient forms of wind converter are the propeller type, for which $0.4 \le C_{pm} \le 0.5$. In addition, the maximum value of the power coefficient is designed to occur in the range of TSR, namely $4 < \text{TSR} < 7$. A more detailed performance characteristic for propeller and Darrieus machines is shown in Fig. 2.8. The peak value of C_p is seen to be approaching 0.4, which is typical of small wind converters.

The power extractable from a freely flowing stream of wind, with a power coefficient $C_p = 0.4$, is shown in Fig. 2.9 using logarithmic scales on both axes.

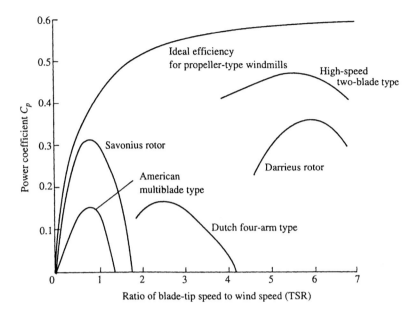

Fig. 2.7 Power coefficient versus tip-speed ratio for various converters.

2.5 Mechanical Features of Wind Machines

2.5.1 *Axial thrust (Pressure)*

The action of the wind stream onto the rotating propeller, as in Fig. 2.3, is to create a pressure force acting along the horizontal shaft, called the thrust, Th. A detailed aerodynamic analysis (not given here) shows that the thrust may be expressed as:

$$\text{Axial Thrust} = \text{Th} = \frac{1}{2}\rho A V_1^2 \left[1 - \left(\frac{V_2}{V_1}\right)^2\right]. \quad (2.31)$$

The thrust may be expressed in terms of the extracted power P_{ex} by comparing Eq. (2.31) with Eq. (2.20):

$$\text{Th} = \frac{P_{ex}}{V_1 \left[\frac{V_1+V_2}{2}\right]} = \frac{P_{ex}}{V_1 V_r}. \quad (2.32)$$

Like the extracted power, the thrust per unit area in the wind stream is determined entirely by the wind velocities. The thrust has to be counteracted by the end bearings on the propeller shaft.

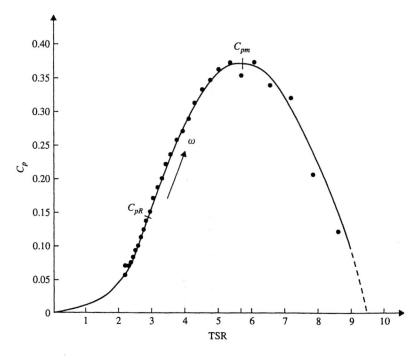

Fig. 2.8 Power coefficient versus tip-speed ratio for Darrieus and propeller machines.[1]

2.5.2 *The "Yaw" effect*

The wind at a given site is subject to rapid and frequent changes of direction. But to maintain efficient operation, the turbine propeller plane must remain perpendicular to the wind direction. This requires that the turbine assembly be free to rotate about a vertical axis—a phenomenon that aeronautical engineers call the "yaw" effect. With good bearings, a machine can be pivoted to swivel under the influence of a vane or a rudder wheel mounted downwind, as illustrated in Fig. 1.4 of Chap. 1.

In large modern wind turbines, a weather vane monitors the wind direction and an electric yaw drive is used to swivel the propeller plane broadside onto the wind.

2.5.3 *Gyroscopic forces and vibrations*

Yawing rotation about the vertical axis while the rotor is turning about its horizontal axis encounters strong gyroscopic forces. These forces have to be

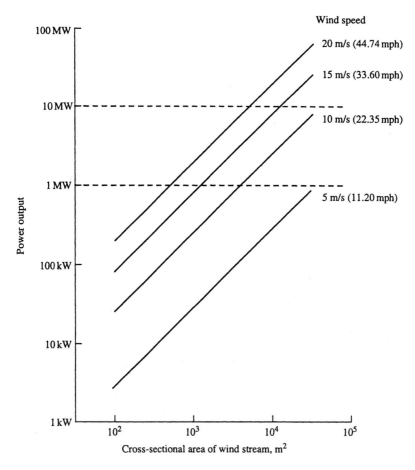

Fig. 2.9 Power extractable from a freely flowing wind stream ($C_p = 0.4$).[1]

transmitted through the bearings and propeller shaft causing high stresses and vibrations. For this reason, the propeller blades of large machines are made of a lightweight material such as a composite plastic such as fibreglass rather than metal.

The action of rotation of the blades results in periodic vibrations. With a downwind-designed machine, which is characteristic of many large systems, each rotating blade passes through the wind shadow of the tower once per rotation. This results in a sudden reduction of air pressure on each blade followed by a sudden increase of air pressure, as it emerges from the shadow of the tower. The result is to apply a bending moment on each blade at its root or hub joint in alternate directions. Continual flexing of the propeller blades at every rotation produces fatigue stresses in the materials. With two-blade propellers, sometimes, the whole rotor is mounted on a single-shaft hinge allowing fore-aft rotation or "teetering" to reduce out-of-plane bending moment fluctuations. In order to minimise the vibration problem, some wind machine designers prefer to use three-blade propellers rather than two-blade propellers, even with the additional cost of the extra blade.

2.5.4 *Centrifugal forces*

The rotation of the blades of a wind turbine causes outward acting centrifugal forces. This phenomenon can be experienced by tying a weight at the end of a string and swinging it around. The outward acting force depends directly on the mass or weight and on the speed of rotation. Calculation of the centrifugal forces on a wind turbine tending to pull the rotating propeller blades out of their sockets is complicated because the weight is distributed non-uniformly along the length of the blade. A simple calculation that assumed all the weight to be concentrated at a fixed radius of rotation would give inaccurate results. In large modern wind turbines, the blades are large and heavy. Moreover, the cost of the blades and propeller unit is a significant portion of the total system cost.

The amount of power taken from the wind at a fixed wind velocity can be adjusted by varying the pitch angle of the propeller blades. This is realised by rotating part of the propeller arms in their sockets, such as adjusting a screw or bolt. In effect, this changes the force and torque exerted on the rotating propeller. The same principle is used in landing a propeller-driven aeroplane to change the thrust on the blades and thereby reduce the speed. The use of the same technique enables the power extracted from a propeller to be kept constant over a range of wind speeds, illustrated in Fig. 2.10. When the wind reaches a maximum acceptable level known as the furling velocity V_f, the pitch angle of the blades can be adjusted so that zero power is extracted. In severe wind conditions, some form of mechanical brake is also applied.

2.5.5 *Solidity factor*

The solidity factor is defined as the total blade area of the rotor divided by the area swept normal to the wind. In a horizontal axis, propeller machine,

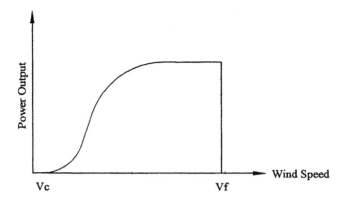

Fig. 2.10 Effect of feathering the propeller.[1] V_c = cut-in speed; V_f = furling speed.

for example, an efficiently designed aerofoil intercepts a large area of wind with a small area of blade. It therefore has a low-solidity factor, which is highly desirable in high-rotational speed systems.

Turbines with high-solidity factor usually suffer from a high degree of aerodynamic interference between the blades, which results in low values of TSR and power coefficient C_p. Examples are the Savonius rotor and the American farm multi-blade type, with the typical performance characteristic given in Fig. 2.7. Wind turbines with high solidity usually operate at low-rotational speeds but have high-starting torques. They are used for direct mechanical applications such as water pumping but are not usually suitable for driving electric generators. For the purpose of electricity generation, it is usual to use low-solidity machines, such as the two-blade propeller, in order to utilise high-operational speeds and high values of power coefficient.

2.5.6 Two rotor blades or three rotor blades?

Most large modern wind turbines are horizontal axis, propeller machines having either two blades (in the USA) or three blades (in Europe) on the rotor. There are long-standing and ongoing differences of view amongst wind and aeronautical engineers as to the merits of the two designs. The maximum achievable values of power coefficient C_p over a range of values of TSR have been calculated under the idealised condition of no aerodynamic drag, for the rotors with several blade numbers.

Within the normal working range of TSR, a three-blade propeller has a slightly larger value (e.g., 5%) of power coefficient. But most two-bladed wind turbines use a higher value of TSR than most three-bladed machines. There is little practical difference in the maximum achievable C_p between two- and three-bladed designs, assuming no drag.

Three-blade machines have the advantage that the polar moment of inertia with respect to yawing is constant, which contributes to smooth operation. A two-bladed rotor has a lower moment of inertia when the blades are vertical than when they are horizontal creating rotation imbalance. An important consideration in selecting the number of blades is that the blade root stress increases with blade number for a turbine of given solidity. In general, increasing the design TSR entails decreasing the number of blades.[5]

2.5.7 Shaft torque and power

Most wind energy systems are used to generate electricity. The wind turbine is usually coupled to a generator directly as in Fig. 2.11 (a) or via a gearbox to step up the generator shaft speed, Fig. 2.11 (b). For this reason, the generator is usually mounted at the top of the supporting tower along with the gearbox. Electric cables run down the tower to connect the generator to its electrical load on the ground below. The torque, speed, and power of a rotating shaft are linked by the relationship, given in Eq. (2.27).

If the torque T is in newton-metres (Nm) and the angular speed of rotation ω is in radians per second then the power is in watts. In Fig. 2.11(b), the shaft torque into the gearbox T_t, from the turbine shaft, is given by:

$$T_t = \frac{P_{ex}}{\omega_t}. \qquad (2.33)$$

Similarly, the torque on the gearbox output shaft, into the generator T_g, is given by:

$$T_g = \frac{P_g}{\omega_g}. \qquad (2.34)$$

Considerable torsional shear stress is imposed on a shaft due to rotational forces. For a solid cylindrical shaft subjected to a torque T, the torsional

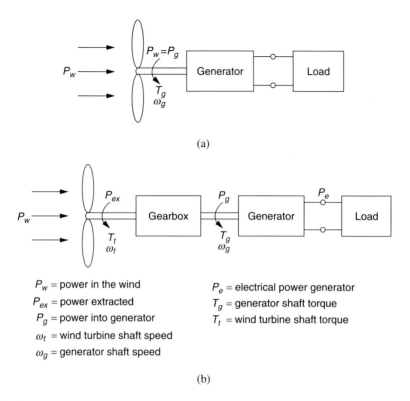

P_w = power in the wind
P_{ex} = power extracted
P_g = power into generator
ω_t = wind turbine shaft speed
ω_g = generator shaft speed

P_e = electrical power generator
T_g = generator shaft torque
T_t = wind turbine shaft torque

(b)

Fig. 2.11 Power train for wind-powered electricity generation: (a) direct-on load, (b) gearbox system.[1]

stress f_s, at any arbitrary radius r_s, Fig. 2.12, is given by:

$$f_s = \frac{T \cdot r_s}{J} \; \frac{N}{m^2}, \qquad (2.35)$$

where J is the polar (area) moment of inertia having the dimension (mass)^4 or m^4. For a solid cylindrical shaft of outer radius r_o, the polar moment of inertia can be shown to be:

$$J = \frac{\pi r_o^4 m^4}{2}. \qquad (2.36)$$

Combining Eqs. (2.35) and (2.36) gives an expression for the shear stress at the surface of a solid cylindrical shaft of radius r_o:

$$f_s = \frac{2T}{\pi r_o^2} \; \frac{N}{m^2}. \qquad (2.37)$$

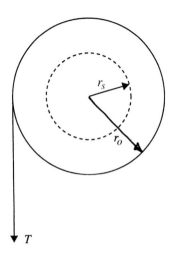

Fig. 2.12 Wind machine shaft.[1]

2.6 Fixed Rotational Speed or Variable Rotational Speed?

The ambient wind in any location is variable for both wind speed and direction. In addition, a turbine is subject to turbulent wind gusts of a transient and unpredictable nature. A design choice has to be made between operating a turbine at variable rotational speed following the wind and regulating the speed of rotation to create a fixed speed or a choice of two (usually) different fixed speeds of rotation. For either option, the turbine must be capable of being completely stalled into total immobility at some predetermined safe maximum operating speed.

Any wind energy system design must aim to optimise the annual energy capture at its given site. In order to operate at its highest efficiency (i.e., with maximum power coefficient C_p), a turbine must operate at its optimum value of TSR as the wind speed varies, as illustrated in Figs. 2.7 and 2.8. The best condition to be aimed for in design is for the turbine to operate, at all wind speeds, at a value of TSR at or close to the value that results in maximum power coefficient. But since the wind speed varies the design issue is therefore either (a) to operate the turbine rotor at a fixed speed by (say) adjusting the pitch angle of the turbine propellers as the wind speed changes or (b) to permit the rotational speed to change, following the variable wind speed.

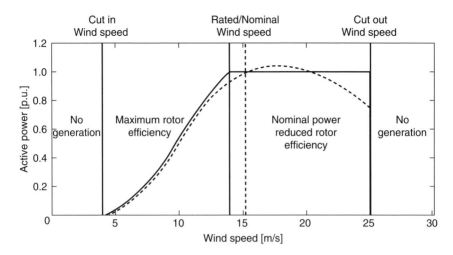

Fig. 2.13 Active power versus wind speed.[6] — pitch controlled; --- stall controlled.

2.6.1 *Constant speed operation*

Operation at a constant speed may be realised by one of the two basic control methods:

1. by varying the pitch angle of the propeller blades as the wind speed varies. This can be achieved by rotating either the whole propeller blades or the tips of the propeller blades in their sockets. This form of control is usually called pitch angle control.
2. by the use of a propeller of fixed pitch angle but where the propeller surfaces are designed to introduce stall over a range of wind speeds. This form of design is usually referred to as "stall regulation" or "stall control".

The two design methods lead to very similar turbine power characteristics, as shown in Fig. 2.13. At low wind speeds (1–3 m/s), the turbines are shut down. Start-up begins at a cut-in speed between 2.5 and 5 m/s. Rated wind speed, at which the nominal output is reached, is in the range of 12–15 m/s. Below the nominal wind speed, the aim is to maximise the turbine rotor efficiency.[6]

When the turbine rotational speed is constant, a coupled A.C. generator will operate at a fixed frequency, which can be synchronised to the frequency of the electrical system to which it is connected. It is not then necessary to use any form of electronic frequency changer, as a decoupling device as is required in variable speed systems. This results in a simpler and cheaper electrical system.

2.6.2 Variable speed operation

From Eq. (2.29), it is seen that to keep the TSR constant, at its desired value, it is necessary to control the turbine rotational speed ω in proportion to the wind speed V. Alternatively, a turbine can be operated at one of the two fixed rotational speeds, which gives a TSR closer to the optimal than a single fixed speed, but is slightly less efficient.

Variable speed control is more expensive than fixed speed control but is found to yield 20%–30% more energy. The additional expense of variable speed control arises from the need for a controlled and variable-pitch propeller and also from the need for a power electronic frequency controller

Table 2.4 Features of fixed speed and variable speed wind-turbine systems.[2]

Fixed speed	Variable speed
Lower capital cost	Operation at optimal TSR results in higher energy capture
Higher reliability	Lower transient torques, because rotor acts as a flywheel
Lower probability of excitation of structural resonances	High performance aerofoil blades with peaky C_p/TSR characteristics
No frequency conversion device needed between the generator and the system	Electrical interface frequency converter required, which can add an ability of damping
Simple and cheap electrical system	Cheaper gearbox but more expensive electrical drive train
Damping system needed for wind speeds above rated (e.g. blade pitch-angle control)	Mechanical damping system not required
Synchronisation of generator to system frequency needed	No synchronisation required
No injection of electrical harmonics from the generator into the supply	Injection of electric current harmonics from the converter into the supply system
	Better quality electrical power. Both active and reactive power elements can be controlled, leading to power factor control

between the variable frequency generator output and the fixed frequency bus bars of the load system. This is discussed in Chap. 7. "The aerodynamic noise generated by a wind turbine is found to be offensive to some people. This noise is approximately proportional to the fifth power of the tip-speed. Both variable speed and two speed operation allow the rotational speed to be significantly reduced at low wind speeds, which reduces the turbine aerodynamic noise greatly".[7]

The cost of generating electricity from a wind-turbine system depends on the efficiency of the energy capture, the capital cost, and the lifetime of the plant and on the reliability of the system.[2] This is discussed in Chap. 10.

A summary of the relative merits of fixed rotational speed operation compared with variable rotational speed operation is given in Table 2.4. Further discussion of many of these various features is given in subsequent chapters.

2.7 Efficiency Considerations of Wind-Powered Electricity Generation

A power flow diagram for a basic wind converter system is given in Fig. 2.11. The combination of turbine, gearbox and generator is sometimes called a power train. The stage efficiencies of the various stages are given by:

$$\text{Turbine efficiency} = \frac{P_{ex}}{P_w} = C_p \tag{2.38}$$

$$\text{Gearbox efficiency} = \frac{P_g}{P_{ex}} = \eta_{gb} \tag{2.39}$$

$$\text{Generator efficiency} = \frac{P_e}{P_g} = \eta_g. \tag{2.40}$$

The overall efficiency η of the three-stage system of Fig. 2.11, from input to output is:

$$\eta = \frac{\text{electrical output power}}{\text{power available in the wind}}$$
$$= \frac{P_e}{P_w} = \frac{P_{ex}}{P_w} \cdot \frac{P_g}{P_{ex}} \cdot \frac{P_e}{P_g}. \tag{2.41}$$

In terms of the individual stage efficiencies, from Eqs. (2.38), (2.39), and (2.40), the overall efficiency η can be written as:

$$\eta = \frac{P_e}{P_w} = C_p \cdot \eta_{gb} \cdot \eta_g. \qquad (2.42)$$

For small wind energy installation, up to a few kW output rating, the overall efficiency η is of the order 20%–25%.

The electrical output power may be written in terms of the total power in the wind P_w and the wind velocity V utilising Eq. (2.42) and substituting from the basic Eq. (2.12).

$$P_e = C_p \cdot \eta_{gb} \cdot \eta_g \cdot P_w = C_p \cdot \eta_{gb} \cdot \eta_g \cdot \frac{1}{2}\rho A V^3. \qquad (2.43)$$

Values of the power coefficient C_p for the two-blade propeller type of the wind turbine vary with size and rating. They are found to be in the ranges shown below.

$$\begin{aligned} C_p &= 0.4 - 0.5 \text{ for large machines } (100\,\text{kW} - 3\,\text{MW}) \\ &= 0.2 - 0.4 \text{ for small machines } (1\,\text{kW} - 100\,\text{kW}) \\ &= 0.35 \text{ typically for small machines.} \end{aligned} \qquad (2.44)$$

Modern types of mechanical gearbox have efficiencies that depend both on the size (rating) and on the speed of rotation. At the rated speeds, it is found that the gearbox efficiencies lie in the following ranges.

$$\begin{aligned} \eta_{gb} &= 80\% - 95\% \text{ for large machines} \\ &= 70\% - 80\% \text{ for small machines.} \end{aligned} \qquad (2.45)$$

Electrical generators have efficiencies that increase with rated size. The process of converting rotational mechanical energy to electrical energy is inherently more efficient than wind-turbine conversion or than any process involving heat to work (i.e., thermodynamic) conversion. For operation at rated output:

$$\begin{aligned} \eta_g &= 80\% - 95\% \text{ for large machines} \\ &= 60\% - 80\% \text{ for small machines.} \end{aligned} \qquad (2.46)$$

It is notable that large systems are more efficient than small systems. In general, an increase of scale is accompanied by an increase of efficiency. Typical values of overall efficiency η can be obtained by substituting typical values of stage efficiencies into Eq. (2.42).

For large machines:
$$\eta = 0.45 \times 0.9 \times 0.9 = 0.36.$$
For small machines:
$$\eta = 0.35 \times 0.75 \times 0.7 = 0.18. \tag{2.47}$$

2.8 Worked Numerical Examples on Wind-Turbine Operation

Example 2.1

Use the data of Fig. 2.7 to predict the approximate diameter, in feet, of a wind stream of circular cross-section that contains an extractable power of 1 MW in a wind speed of 38 mph, $C_p = 0.4$.

_._._._._._._._._._._._._._._._._._._._

The 1 MW coordinate in Fig. 2.7 intersects the line representing 15 m/s (33.6 mph) at a cross-sectional area of approximately 1100 m². Interpolating between coordinates on a non-graduated logarithmic scale requires careful attention. The value 38 is almost midway between the calibrated characteristic parameters of 44.74 and 33.6 in Fig. 2.7 but the 38 mph characteristic does not lie midway between the two calibrated characteristics.

By estimation, 1 MW extractable power, at 38 mph, corresponds to a wind-stream area of 1,000 m²

$$A = 1000 \, \text{m}^2$$
$$= 1000 \times \left(\frac{39.37}{12}\right)^2 = 10764 \, \text{ft}^2.$$

But $A = \frac{\pi D^2}{4}$

$$D^2 = \frac{4 \times 10764}{\pi} = 12732 \, \text{ft}^2$$
$$D = 112.84 \, \text{ft}.$$

Example 2.2

The largest wind turbine in the British Isles to date (2010) is the 3-MW generator system in the Orkney Isles, north of Scotland. Use the data of

Fig. 2.7 to verify roughly its swept diameter for operation in a 38-mph wind.

_._._._._._._._._._._._._._._._._._._

The Orkney wind-generator system is rated 3-MW electrical output power. To realise this, the extractable power from the wind P_{ex} would need to be (say) 20% or 0.75 MW bigger. Interpolating between the 1-MW and 10-MW gradations in Fig. 2.7, for 3.75 MW, is about one-half way and corresponds to a horizontal intercept of roughly A = 3000 m². The actual design diameter of the Orkney machine is 60 m, giving a swept area:

$$A = \frac{\pi D^2}{4} = \frac{\pi \times 3600}{4} = 2827.4 \, m^2.$$

Despite the limitations of reading from a small-scale logarithmic data sheet, it can be seen that there is good correlation.

Example 2.3

A wind turbine of the two-blade propeller type is designed to have its maximum power coefficient value at a TSR = 6, when the wind velocity is 25 mph. If the blade diameter is 100 ft, what is the recommended speed of rotation?

_._._._._._._._._._._._._._._._._._._

$$V = 25 \, mph = 25 \times 0.447 = 11.18 \, m/s.$$

From Eq. (2.28):
  with TSR = 6

$$v = r\omega = V \times TSR = 6 \times 11.18 = 67.1 \, m/s$$

$$D = 100 \, ft = 100 \times \frac{12}{39.37} = 30.5 \, m$$

$$r = \frac{D}{2} = 15.25 \, m.$$

Speed of rotation:

$$\omega = \frac{v}{r} = \frac{67.1}{15.25} = 4.4 \, rad/s.$$

From Eq. (2.30):

$$n = \frac{60\omega}{2\pi} = 42 \, rpm.$$

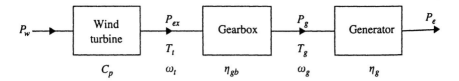

Fig. 2.14 Power flow in a wind converter system.[1]

Example 2.4

Evaluate typical values of overall working efficiency for (a) large wind systems and (b) small wind machine systems, assuming realistic values of the various stage efficiencies.

_.

(a) For large wind-turbine systems, typical stage efficiencies are:

from Eq. (2.44) $C_p = 0.42$
from Eq. (2.45) $\eta_{gb} = 0.85$
from Eq. (2.46) $\eta_g = 0.92$.
Then, from Eq. (2.42), $\eta = 0.42 \times 0.85 \times 0.92 = 33\%$.

(b) For small wind-turbine systems, typical stage efficiencies are:

from Eq. (2.44) $C_p = 0.3$
from Eq. (2.45) $\eta_{gb} = 0.75$
from Eq. (2.46) $\eta_g = 0.7$
from Eq. (2.42) $\eta = 0.3 \times 0.75 \times 0.7 = 16\%$.

There is seen to be an overall efficiency advantage of the order 2:1 in using large-scale wind generation.

Example 2.5

Wind-turbine units are rated at 2 MW in a rated wind speed of 13 m/s. The stage efficiencies are $C_p = 0.32$, $\eta_{gb} = 0.94$, $\eta_g = 0.96$. What is the necessary swept area? If the rotor is a two-blade propeller (horizontal axis), what is the diameter? ($\rho = 1.29 \text{ kg/m}^3$.)

$$\eta = 0.32 \times 0.94 \times 0.96 = 0.29$$
$$P_e = 2 \text{ MW}$$
$$\therefore P_w = \frac{P_e}{\eta} = \frac{2 \times 10^6}{0.29} = 6.9 \times 10^6 \text{ W}.$$

Now:

$$P_w = \frac{1}{2}\rho A V^3$$

$$6.9 \times 10^6 = \frac{1}{2} \times 1.29 \times A \times (13)^3$$

$$A = \frac{2 \times 6.9 \times 10^6}{1.29 \times 13^3} = 4870 \, \text{m}^2.$$

For a circular area:

$$A = \frac{\pi D^2}{4}, \quad D = 78.8 \, \text{m} \, (258 \, \text{ft}).$$

The comparatively large diameter is because of the low value of the turbine power coefficient.

Example 2.6

A generator driven by a wind turbine is required to deliver 1 MW of power at the generator terminals. The turbine is a two-blade propeller rotating about a horizontal axis and the maximum permitted shear stress of the turbine shaft is $55 \times 10^6 \, \text{N/m}^2$. The rotor is designed to operate at a rotational speed of 22 rpm.

(a) If the turbine delivers its rated power at a wind average speed of 25 mph, calculate the corresponding diameter of the propeller and its TSR, assuming a typical value for the overall efficiency. The air density may be assumed to have a value $1.29 \, \text{kg/m}^3$.
(b) Calculate the torque on the turbine shaft and the necessary shaft diameter.

_._._._._._._._._._._._._._._._._._._

(a) $P_{\text{elect}} = 1.0 \, \text{MW}$
Let $\eta = 0.35$ overall

$$\therefore P_{\text{wind}} = \frac{P_{\text{elect}}}{\eta} = \frac{1.0 \times 10^6}{0.35} = 2.86 \times 10^6 \, \text{W}.$$

Now $25 \, \text{mph} \equiv 11.175 \, \text{m/s}.$

But, wind power $P_{\text{wind}} = \frac{1}{2}\rho A V^3$

$$\therefore A = \frac{2 \times 2.86 \times 10^6}{1.29 \times (11.175)^3} = 3171.8\,\text{m}^2.$$

But, $A = \frac{\pi D^2}{4}$.

$$\therefore D = \sqrt{\frac{4 \times 3172}{\pi}} = \sqrt{4038.7}$$
$$= 63.55\,\text{m}\ (208.51\,\text{ft}).$$

$$\text{TSR} = \frac{r\omega}{V}$$

$$r = \frac{D}{2} = 31.38\,\text{m}$$

$$\omega = 22 \times \frac{2\pi}{60} = 2.3\,\text{rad/s}.$$

$$\therefore \text{TSR} = \frac{31.78 \times 2.3}{11.175} = 6.54.$$

(b) $T = \dfrac{P}{\omega} = \dfrac{2.86 \times 10^6}{2.3} = 1.24 \times 10^6\,\text{Nm}$

$$f_s = \frac{Tr_s}{J} = \frac{Tr_s}{\frac{\pi r^4}{2}} = \frac{2T}{\pi r_s^3}, \text{ where } r_s = \text{shaft radius}$$

Now stress

$$\therefore r_s^3 = \frac{2T}{\pi f_s} = \frac{2 \times 1.24 \times 10^6}{\pi \times 55 \times 10^6} = 0.01\,\text{m}^3$$

$$\therefore r_s^3 = \frac{10}{1000}$$

$$r_s = \frac{\sqrt[3]{10}}{10} = 0.215\,\text{m}\ (8.46\,\text{inches}).$$

Example 2.7

The Orkney wind machine in Great Britain is rated at 3 MW, at the rated wind speed of 17 m/s, with a blade diameter of 60 m. What is the power coefficient of the turbine? ($\rho = 1.29$ kg/m^2.)

From Eq. (2.12),
$$P_w = \frac{1}{2}\rho A V^3$$
$$= \frac{1}{2} \times 1.29 \times \frac{\pi 60^2}{4} \times 17^3 = 8959810 \text{ W}$$
$$P_e = 3 \text{ MW}.$$

Overall efficiency is then, from Eq. (2.42):
$$\eta = \frac{3 \times 10^6}{8959810} = 0.335 \text{ p.u.}$$

But, $\eta = C_p \cdot \eta_{gb} \cdot \eta_g$.
If $\eta_{gb} = 0.90$ and $\eta_g = 0.90$, the turbine power coefficient is:
$$C_p = \frac{0.335}{0.9 \times 0.9} = 0.414.$$

2.9 Problems and Review Questions[1]

2.1 Use the information in Eq. (2.20) to show that maximum power is extracted from a wind stream when the upstream velocity V_1 is three times the downstream velocity V_2.

2.2 If the wind speed in a certain location is doubled, how does this affect the power output of a wind generator?

2.3 Use the characteristics of Fig. 2.9 to estimate the necessary swept diameter of a large wind turbine to generate 10 MW of electrical power in a 25-mph wind.

2.4 Explain, using a diagram, the term "tip-speed ratio" (TSR). If the optimum TSR $= 6.0$ for a two-blade propeller of diameter 180 ft, what speed of rotation in rpm must be used in a 20-mph wind?

2.5 A large propeller-type wind turbine has a diameter of 200 ft. If the speed of rotation at full load is regulated to 32 rpm when the wind speed is 30 mph, what is the value of the TSR?

2.6 What are the extreme limits of overall efficiencies for (a) large wind-turbine systems and (b) small wind-turbine systems, indicated by the stage efficiency values of Sec. 2.7?

2.7 It is required to generate 1200 W of electrical power at the terminals of a generator driven through a gearbox by a wind machine. The location is such that the wind supply is of smooth laminar flow with an average speed of 17.5 mph. Assign typical efficiency values to the components of the system and calculate the blade diameter required for a good quality two-blade propeller type of wind machine (air density = 1.29 kg/m^3).

2.8 A two-blade propeller wind turbine is coupled to a 10-kW electric generator. It is desired to generate 10 kW at the load terminals. The average wind speed is 8 m/s and can be considered as ideally smooth. Assign typical values of efficiency to the turbine and the generator and calculate the necessary diameter of the area swept out by the rotating blades (air density = 1.29 kg/m^3).

2.9 A two-blade propeller is used as a wind turbine directly on the shaft of a small electric generator. Assign typical efficiencies to the wind turbine and the generator and calculate the blade diameter required to generate 500 W at the load terminals in a wind of average speed 15 mph ($p = 1.29$ kg/m^2).

2.10 A two-blade propeller wind machine has a blade diameter 3.5 m and a power coefficient $C_p = 0.36$. What average wind speed in mph would result in 1 kW of power generation if the generator has 70% efficiency?

2.11 (a) A two-blade propeller wind turbine has a blade diameter of 4 m. What average wind speed would result in a power output of 500 W at the generator terminals, assuming typical efficiencies for the generator, the gearbox, and the turbine?

(b) At your calculated wind speed, what is the TSR if the turbine shaft speed is 40 rpm?

(c) Calculate the diameter of the turbine shaft if the maximum permitted shear stress is 55×10^6 N/m^2.

2.12 A wind-turbine system consists of a two-blade propeller, rotating about a horizontal axis, driving an electrical generator via a gearbox. The generator is required to deliver its rated capacity of 2 MW with the wind turbine rotating at 18 rpm. The maximum permitted shear tress of the solid steel turbine shaft is 55×10^6 N/m^2. The design estimate for the overall efficiency is 30%.

(a) Calculate the torque on the turbine shaft at rated load and the necessary shaft diameter.
(b) Define the term "tip-speed ratio" (TSR) and explain its use in wind energy calculations.
(c) Calculate values for the TSR and the required diameter of the propeller if the rated turbine power is developed in an average wind speed of 27.5 mph (air density = 1.29 kg/m^3).
(d) Estimate realistic and consistent figures for the turbine power coefficient, the gearbox efficiency, and the generator efficiency.

2.13 A solid-steel cylindrical turbine shaft has a diameter of 12 inches. If the shear stress coefficient $f_s = 55 \times 10^6$ N/m^2, what is the maximum permitted shaft torque?

2.14 The maximum safe rotational speed for a certain wind turbine is 35 rpm. How is this maximum speed retained in high winds? What problems would arise if the speed of rotation became excessive? What mechanism is used on the cupola of a Dutch-type windmill to keep the rotating blades facing upstream?

2.15 The jet stream over the North Atlantic Ocean travels from west to east at 100 mph. How does this affect (i) the air speed and (ii) the ground speed of 500 mph high-flying jetliners crossing the ocean between Europe and the USA?

2.16 On a certain day, the North Atlantic jet stream travels from west to east at 95 mph. If the average air speed of a jetliner is 445 mph, how long the flight will take to travel from Manchester, UK to Atlanta, a distance of 3400 miles?

References

1. Shepherd, W., D. W. Shepherd, *Energy Studies*, Second Edition, Chapter 10, Imperial College Press, London, England, 2003.
2. Freris, L. L., *Wind Energy Conversion Systems*, Prentice Hall International (UK) Ltd., Herts, HP2 4RG, England, 1990.
3. Golding, E. W., *The Generation of Electricity by Windpower*, E. and F.N. Spon Ltd., London, England, 1955.
4. Patel, M. R., *Wind and Solar Power Systems*, CRC Press, New York, USA, 1999.

5. Manwell, J. F., J. G. McGowan, A. L. Rogers, *Wind Energy Explained: Theory, Design and Applications*, John Wiley & Sons Ltd., Chichester, England, 2002.
6. Slootweg, H., E. de Vries, Inside Wind Turbines, *Renewable Energy World*, **6**(1), 31–40, 2003.
7. Burton, T., D. Sharpe, N. Jenkins, E. Bossanyi, *Wind Energy Handbook*, John Wiley & Sons Ltd., Chichester, England, 2001.

CHAPTER 3
Past and Present Wind-Energy Turbines

3.1 Nineteenth-Century Windmills

The use of wind power, mainly for the pumping of water, was very extensive in Europe and North America during the nineteenth century. It is estimated that, since the mid-nineteenth century, more than six million machines of the American farm types, shown in Fig. 3.1, have been used in the USA. The metal-blade rotor is roughly 12–16 ft diameter, with a tail-vane stabilizer to keep the multi-sails upstream and is ideal for remote farm locations. Each individual turbine is rated at less than one horsepower (HP). Typically for this type of machine, the solidity factor is high, being of the order 0.7.

By 1850, the use of windmills in the USA provided about 109 kWh of energy per year.[1,2] The installation of steam engines began to support the use of wind converters in the later nineteenth century. But the use of this type of small system continued into the 1930s in the USA, both for pumping water and to power small electric generators, rated up to 2 kW, used for charging lead-acid storage batteries. These, in turn, provided power for the lighting of barns and animal sheds.

In Europe, corresponding developments took place in several countries, notably Denmark, Germany, Holland, Russia, and Great Britain.

Fig. 3.1 American Farm windmill used for pumping water. [http://www.windmills.net/10windmill.jpg]

For example, in Denmark, by the end of the nineteenth century, about 2500 industrial windmills were in operation, having a total rating of about 40,000 HP (30 MW). Most of these were in rural areas. Also, there were approximately 4600 windmills on Danish farms, used for threshing and milling grain and for the pumping of water. By the 1930s, industrial applications were much reduced but the number of small farm units had increased to about 16,000.[1] Denmark has always been a world leader in the design and use of wind-energy systems and remains so to the present day.

3.2 Early Twentieth-Century Wind-Energy Turbines

Towards the end of the nineteenth century and into the twentieth century, the application of wind energy changed from milling grain and water pumping to the generation of electricity. This required the development of turbines of low solidity factor and much higher speeds than were realisable by multi-blade farm windmills.

The first use of a large windmill to generate electricity in the USA was a system built in Cleveland, Ohio in 1888 by Charles F. Brush. Brush's machine had a multi-bladed rotor, 17 metres in diameter, with a large fan tail. It was the first wind machine to incorporate a step-up gearbox (gear ratio 50:1) and drove a direct-current (DC) generator at 500 rpm. The generator was rated at 12 kW and operated for 20 years.[3] By 1920, the two dominant rotor configurations, fan-type and sail-type, had both been found to be inadequate for generating appreciable amounts of electricity. The further development of wind-powered electrical systems in the USA was inspired by the design of aeroplane propellers and aerofoil wing sections.[3]

By the mid-1920s, small (1–3 kW) wind-powered DC generators were in widespread use in rural areas, driven by low-maintenance three-blade propeller turbines mounted on tall (e.g., 70 ft) towers (Fig. 3.2). The predominant companies in the USA were Parris-Dunn and Jacobs Wind-Electric. These were gradually forced out of business by the customer demand for larger amounts of grid (utility) supplied electricity. The escalating price of oil in the latter years of the twentieth century has caused a comeback and modern successors of the early machines can now be seen all over the country.

Fig. 3.2 Propeller driven wind turbine for generating electricity [Photographed by author WS in Wisconsin, USA-1997].

In 1891, Professor Poul La Cour in Denmark developed the world first wind-powered electricity generating system to employ aerodynamic design principles. The machines incorporated low-solidity, four-blade rotors 75 ft in diameter that were designed with primitive aerofoil sections.[3] By about 1920, the use of 25-kW, high-speed wind generators was common in Denmark. As in America, however, cheaper fossil-fuel steam plants put the wind-powered electricity generation industry out of business for many years.

Large-scale wind-energy conversion systems based on aerodynamic designs were first undertaken in Russia in 1931. The 100-kW wind turbine had a generator and controls mounted at the top of a 100-ft tower (Fig. 3.3). The heel of the inclined strut was mounted on a carriage that ran on a circular track to keep the rotor facing into the wind.[1] This turbine was connected by a 6300-V transmission line to a 20-MW steam-powered station 20 miles away. Subsequent experimental large-scale systems in Denmark,

Fig. 3.3   Russian wind-turbine; 100 kW (24.6 mph rated wind speed).

France, Germany, Great Britain, and the USA all failed to result in practical economic designs.

The first successful large wind turbine was the Smith-Putnam machine built at Grandpa's Knob, Vermont, USA in 1941.[4] This privately funded venture proved to be the prototype and inspiration for what has now become a vast industry. As with all horizontal axis propeller machines, the generator and gearbox were mounted on the turbine shaft in a housing at the top of the tower (Fig. 3.4). The Smith-Putnam machine was a two-blade propeller downstream system, rated at 1.25 MW, with a blade diameter of 53.34 m (175 ft). It operated at a constant rotational speed of 28 rpm. For 35 years, it held the record as the world's largest wind machine system. The electric generator was a synchronous machine that fed electrical power directly into the Central Vermont Public Service Corporation electricity grid.

The Smith-Putnam machine suffered two mechanical failures. After a main bearing replacement, there was a spar failure causing one of the propeller blades to fly off. The operating company decided that a repair would be uneconomical and the venture was closed down in 1945. Although the

46 *Electricity Generation Using Wind Power*

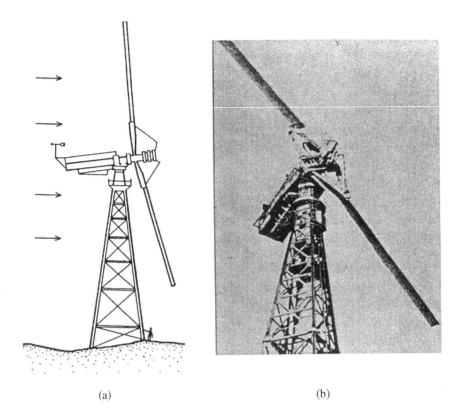

(a) (b)

Fig. 3.4 Smith-Putnam wind machine[4] (a) diagrammatic, (b) on-site photograph.

Smith-Putnam machine operated for only 18 months in all, it was a proving ground for high-power wind generation systems. The mechanical failures were due to the limitations of the knowledge of the materials available at the time and not to the basic system design. Better materials are now available. Also, the engineering knowledge about the bearing design and about the fatigue failure of metals make it unlikely that the Smith-Putnam type of failure would occur in modern wind turbines. Although the Smith-Putnam project was superficially a failure, it worked well for long enough to prove that large-scale, wind-powered electricity was feasible. All of the principal engineering challenges were overcome opening the way to future developments.

Considerable work was done on wind-powered electricity generation in Great Britain in the late 1940s and the 1950s.[5] For example, in 1950,

Fig. 3.5 British wind-turbine, John Brown design located on Cape Costa in the Orkney Islands; 100 kW (35 mph rated wind speed).

the North Scotland Hydroelectric Board commissioned an experimental 100-kW, three-blade machine (Fig. 3.5). This operated for short periods in 1955 coupled to a diesel-powered electricity network but had to be shut down because of operational problems.[1]

In the 1950s, the Enfield Electric Cable Company built a unique 100-kW wind-powered generator (Fig. 3.6), at St. Albans, England. The rather dramatic aeronautical-looking design was due to a Frenchman named Andreau. This consisted of a hollow tower of 85 ft high and a hollow rotor with openings at the blade tips. A pressure differential drove air from the openings near the base of the tower, up the tower, through an air turbine and out of the rotor blade tips. The design proved to be of low efficiency and was moved to the windier coast of North Wales. Because of the local environmental objections, it was then sold on to Algeria, where it operated successfully for a number of years.

48 *Electricity Generation Using Wind Power*

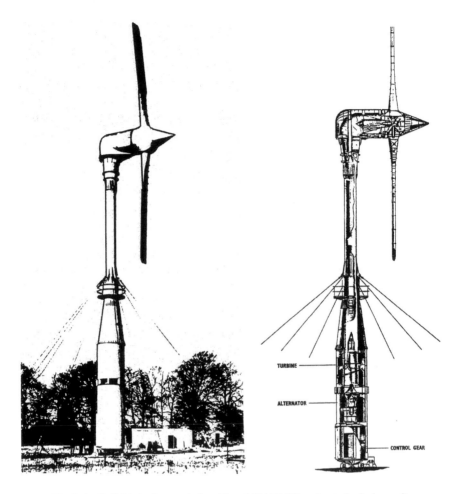

Fig. 3.6   Enfield-Andreau wind-turbine; 100 kW (33 mph rated wind speed).

3.3 Later Twentieth-Century Wind-Energy Turbines

After the Second World War, through the 1950s, a series of experimental wind-energy turbines were developed in Denmark. The latest in the series was the 200-kW Gedser mill (Fig. 3.7), which was a horizontal-axis three-blade propeller machine. This operated until 1968 but was then closed down because it was uneconomic compared with fossil-fuel-powered steam plants. The energy crisis precipitated by the sudden increases in the price of

Past and Present Wind-Energy Turbines 49

Fig. 3.7 Restored Gedser wind turbine 200 kW at 33.6 mph. (Ref. Danish Wind Industry Association, 2009) (http://www.talentfactory.dk/en/pictures/juul.htm).

oil in 1973 caused a change of policy so that the Gedser mill was refurbished and reinstated in operation in 1977.[1]

Several three-blade propeller designs of wind-operated electric generators were developed in France during the period 1958–1966. One unit, near Paris, was rated at 800 kW, with a rotor of 100 ft diameter mounted on a 100-ft tower. This operated at the constant speed of 47 rpm with its synchronous alternator, operating at 1,000 rpm and 3,000 V, connected to a 50-Hz and 60-kV electricity grid via a step-up transformer and a 15-km transmission line. A further unit, in Southern France, operated at 56 rpm with a 70-ft diameter rotor. This used an asynchronous generator with a nominal speed of 1,530 rpm.[1]

In Germany, Professor Ulrich Hutter developed a series of advanced horizontal-axis designs that utilised aerofoil-type fibreglass and plastic blades with variable pitch, using diameters as large as 110 ft. The two-blade

Fig. 3.8 Hutter's wind turbine mid-twentieth century.[3]

propeller design (Fig. 3.8) sought to reduce bearing and structural failures by the techniques of load shedding. One of the most innovative of such schemes was the use of a hinge bearing at the rotor hub that allowed the rotor to "teeter" in response to wind gusts and vertical wind shear. The German design features were later used by US designers.[3]

The issue as to whether to use two blades or three blades in wind turbine design has always been contentious. Some of the relevant technical considerations are discussed in Sec. 2.5.6 of Chap. 2. Any claim about the superiority of one choice is always refuted by the proponents of the alternative. There does not appear to be any definitive scientific basis for the choice. The use of two blades represents a saving on first costs because the rotor blades are the expensive items of any design. But it is claimed that turbines with three blades run smoother, certainly on small machines.[6] Aerodynamic studies indicate that, under ideal conditions, the use of three-propeller blades, rather than two, results in slightly higher values of power coefficient at typical values of tip speed ratio.[7]

3.4 Modern Large Wind Power Installations

The design construction and operation of large (i.e., a few megawatts (MW) down to hundreds of kilowatts (kW)) wind turbine systems for electricity generation is now a well-established technology. Most large-scale systems use horizontal axis and two-blade propeller designs but three-blade systems are also widely used. Although the basic scientific and engineering problems of turbine design and location have been solved there still remain major problems arising from political, environmental, and economic considerations.

A crisis in the price and world supply of oil from the Middle Eastern oil-producing countries (OPEC) in the early 1970s was a great incentive to the industrialised countries of Western Europe and North America to investigate alternative supplies of energy. Since the early 1970s, the USA, in particular, has made a massive investment of effort in the development of wind turbines and wind-powered electricity generation schemes. This has been organised via the US Electrical Development Association (ERDA).

In 1974, the ERDA Model Zero (MOD-0), 100 kW, machine was reported.[8] This was a two-blade variable-pitch propeller, downstream turbine with a diameter of 38.1 m (125 ft) (Fig. 3.9). The AC synchronous generator was driven via a step-up gearbox and fed electrical power directly into the local grid. Subsequent MOD-0 machines were upgraded to 200-kW rating.

The MOD-1 wind turbine system commissioned in 1979 became, at the time, the world's largest machine. This was a two-blade propeller, downwind machine, rated at 2 MW, with a blade diameter of 60.96 m (200 ft). The MOD-1 design was followed in 1980 by the first MOD-2 machine, rated at 2.5 MW, with a blade diameter of 91 m (298.6 ft) and upstream orientation (Fig. 3.10). A group of three MOD-2 machines, on a fixed site, with a hub height of 60 m and providing a test bed for examining the effects of machine clustering, was operating in 1984. A MOD-5 programme involving two-blade, upwind machines rated at 7.3 MW, with a blade diameter of 122 m (400 ft), was planned for the late 1980s. This plan was abandoned, not for technical reasons but because the reduced price of oil and the reduced demand for electricity made it then uneconomical.[9]

Fig. 3.9 ERDA-NASA experimental wind-turbine generator — 100 kW test bed.

Large wind machines have a cut-in speed, V_c (Fig. 2.8), of about 10 mph and rotate at speeds upto about 100 rpm. For economic reasons, large systems need to operate in locations where the average wind speed is greater than 15 mph.

Wind turbine design and development in Europe are dominated by Denmark and Germany. During the 1980s, work concentrated on turbines with ratings of several hundreds of kW. These are now being upscaled to 800–900 kW. The first generation of 1.5-MW machines in 1995–1996 was successful and by the year 2002, there were 1,100 machines of this class in operation. A total of 1,496 turbines of average rating of 1,150 kW were installed in Germany during the year 2000. The largest wind farm in Europe is an array of offshore 2-MW, three-blade propeller turbines at Middelgrunden, Denmark.[10] In 2008, Denmark obtained 20% of its electricity from the wind and is also the world's leading manufacturer and exporter of wind turbines.[11]

Fig. 3.10 American NASA MOD-2 machine 2.5 MW, 2-blade propeller.

The largest wind turbine now (2010) operating, very successfully, in the British Isles is situated in Orkney. Located off the north coast of Scotland, the Orkney Islands are in one of the windiest locations in the world, with average wind speeds of the order 17 m/s (17/0.447 = 38 mph). The high wind speed permits the 3-MW upstream machine to use the relatively small blade diameter of about 60 m (197 ft) at a hub height of about 46 m (151 ft). Electrical power generated by the Orkney machine (Fig. 3.11) is used in the islands and replaces the expensive diesel-electric generation previously used.[12] An impression of the size and internal complexity can be gauged from (Fig. 3.12).[12] Wind speeds up to 60 mph can be utilised. The excessively windy location means that the machine is expected to generate more electrical energy (up to 9,000 MWh/year) than any other known existing wind installation. The physical scale of modern turbines is illustrated in Fig. 3.13.[13]

Fig. 3.11   3 MW wind machine in Orkney, Scotland.[12]

An assessment of the necessary scale of the wind generation can be made by comparing the output power, 3 MW, of modern high-efficiency turbines with a conventional large power station that produces 1,000–2,000 MW. To replace the electrical output power of a 1000-MW fossil fuel or nuclear station would require 333 Orkney size machines or 400 ERDA MOD-2 machines. To take account of the necessary spacing between large wind turbines in the same cluster (i.e., approximately 1,500 ft for 2-MW machines), 1,000 MW of generation could require as much as 500 square miles of ground site. This could create local ecological disruption and aesthetic objections. In the USA, the necessary tracts of land are available in the area of the windy western Great Plains. In the much smaller countries of Western Europe, land is scarcer, more expensive, and in demand for other uses such as farming. The land area required for wind-generated electricity can also be used if it is suitable for the farming of crops or grazing of animals.

In April 2001, the UK government accepted bids for sites designed to produce 1,500 MW of wind-powered generating capacity. Further in

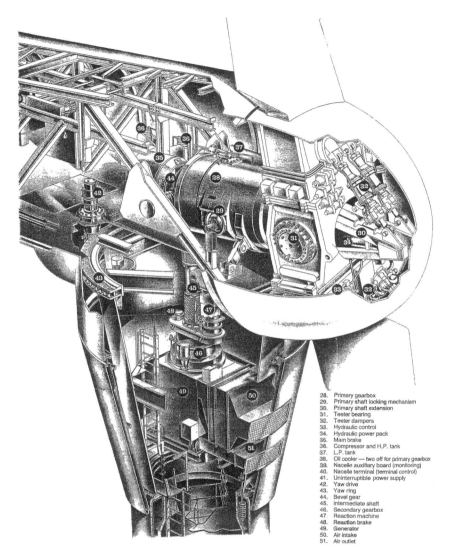

Fig. 3.12 Artistic impression of the internal construction of a 3 MW wind machine.[12]

28. Primary gearbox
29. Primary shaft locking mechanism
30. Primary shaft extension
31. Teeter bearing
32. Teeter dampers
33. Hydraulic control
34. Hydraulic power pack
35. Main brake
36. Compressor and H.P. tank
37. L.P. tank
38. Oil cooler — two off for primary gearbox
39. Nacelle auxiliary board (monitoring)
40. Nacelle terminal (terminal control)
41. Uninterruptible power supply
42. Yaw drive
43. Yaw ring
44. Bevel gear
45. Intermediate shaft
46. Secondary gearbox
47. Reaction machine
48. Reaction brake
49. Generator
50. Air intake
51. Air outlet

December 2003, bids were received for 15 additional offshore sites with a generating capacity of 7000 MW. The total project is intended to satisfy the residential electrical power requirements of 10 million people.[11]

The largest onshore wind farm in Europe in 2009 was the Whitelee Wind farm at Eaglesham, Renfrewshire, Scotland, which is a few miles south of

Fig. 3.13   Bouin wind farm, France[13] N80/2,500 kW turbines.

Glasgow. This consists of 140 Siemens SWT turbines, each rated at 2.3 MW, with a total site rating of 322 MW. The three-blade, downwind fibreglass rotors have a radius of 40 m and are mounted on tapered tubular steel towers to operate at a hub height of 100 m. Operation began in 2009 and the yearly energy production is expected to be 322 GWh. A photograph of part of the farm is given as Fig. 3.17. This wind farm has a very good visitor center with a lot of working models. The installation has authorisation for a further 36 turbines, which is in line with the Scottish government target of generating 31% of Scotland's electricity from renewable energy by 2011 and 50% by 2020.

Most of the 28,400 MW of generation capacity on land now has been developed in Europe and much exploration is taking place of various offshore sites. For example, the Nysted wind farm in Denmark (Fig. 3.14), consists of 72 Bonus turbines each of 2.3 MW capacity.[14] Also, the Arklow

Fig. 3.14 The 165.6 MW Nysted offshore wind farm, Denmark, using 72 Bonus turbines, opened at the end of 2003. How will the larger wind farms of the future meet the challenges of interconnection and energy storage? NYSTED HAVMOELLEPARK.[14]

Bank wind farm, of the Irish coast (Fig. 3.15), uses turbines of greater capacity than 3 MW.[15] As the capacity of the turbines continues to increase, the challenges of interconnection and energy storage will become more demanding.

The modern wind-generation industry, with its aeronautical rotor designs, was born in California, the USA during the early 1980s. But the USA, with 6,370 MW of wind generating capacity, has now (2009) fallen behind Europe. When the US Department of Energy (DOE) released its first wind-energy resource inventory in 1991, it pointed out that three wind-rich states — North Dakota, Kansas, and Texas — had enough harnessable wind energy to satisfy the whole national electricity needs.

Advances in wind turbine design since then enable turbines to operate at lower wind speeds, which increases the efficiency of conversion and also enlarges the regime of wind capture. In 2004, new wind turbine towers were 100 m tall, with much longer blades than the designs of 10 years earlier. This

Fig. 3.15 Big ambitions will need big turbines — Arklow Bank, off the Irish coast, is now being completed. So far, it is the only offshore project using turbines over 3 MW.[15]

results in approximately tripling the amount of harvestable wind. Since the early 1980s wind-generated electricity in the USA has reduced in cost from $0.38/kWh to $0.04/kWh in the prime wind states.[11]

The USA is the world's greatest economic and military power. Its economy is greatly tied to a carbon-dominated energy market and particularly to the price of oil, of which it is the world largest per-capita consumer.[16] The USA has the greatest installed capacity of wind generation, as shown in the Table 8.1.[17] But several other countries now have greater wind utilisation growth rates than the USA. The very large land area of the USA includes many sites suitable for large wind farms, such as the inaptly named Carbon County, Wyoming (Fig. 3.16).[18] It remains an interesting question as to how the use of the renewable forms of energy, notably wind energy, will advance in competition with the large and traditional carbon-based industries of coal, oil, and natural gas.

Fig. 3.16 Rock River Wind Farm, Carbon Country. Wyoming: renewables still exist in a carbon-dominated marked environment SHELL RENEWABLES.[18]

Fig. 3.17 Whitelee Wind Farm, Eaglesham, Scotland, 140, 2.3 MW, 3-blade cantilevered construction (courtesy Scottish Power).

3.5 Worked Numerical Example

Example 3.1

A report by the Electrical Research Association of England suggests that there are about 1500 UK land-based sites, having wind speeds of 20 mph, suitable for wind turbine-generator systems. What portion of the UK demand for electricity could be supplied?[19]

Assume that one could mount one 3-MW turbine or three 1-MW turbines in each location. If all of the turbines in all the locations were working to capacity at the same time — an unlikely eventuality — the total power available would be:

$$P_{total} = 3 \times 1500 = 4500 \text{ MW}.$$

This is equivalent to the power output of four or five medium–large fossil power stations. There are 8760 hours/year. If all of the wind systems operated to capacity for 24 hrs/day and 365 days/year, the energy produced would be:

$$W_{total} = 365 \times 24 \times 4500 = 39.42 \times 10^6 \text{ MWh/yr}.$$

If one includes consideration of downtime for repair and maintenance and also periods of slack wind, then about one-third of this might be available, amounting to about 13 TWh/year.

In 2008, the total domestic UK electricity consumption was 117.8 TWh, which was 34% of the total consumption. The reported land-based wind supply would therefore contribute 13/35.05 or 3.7% of the total demand. This compares with the official figure of the proportion of electricity generated by "renewable and waste", which was 4.2%.[21]

3.6 Vertical Axis Wind Machines

Most of the earliest historical accounts of wind machines refer to vertical axis structures used for pumping water. Small vertical axis machines with rotating vanes or rotating cups are now very widely used for instrumentation purposes, such as wind measurement. Currently, research is being devoted to vertical axis wind turbine systems for electricity generation, especially in the low- and medium-power ranges (a few tens of watts up to tens of kilowatts). The effective rotor surfaces move in the wind direction, rather than perpendicular to it. It is a feature of vertical axis machines that they accept the wind forces equally well from any direction. The issues of upstream, downstream, tower shadow, and yaw that occur in horizontal axis propellers do not arise. The orientation of the blades into the wind

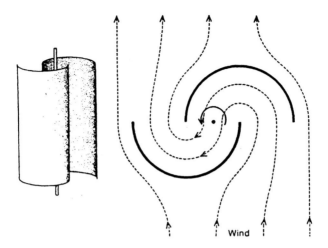

Fig. 3.18 Savonius rotor and its air-stream flow.[24]

is not required. Turbine power coefficients tend to be low — usually less than 1/3 — that is the theoretical maximum. With a vertical axis machine the electric generator can be mounted on the ground at the bottom of the shaft.[22,23]

3.6.1 *The Savonius design*

The most basic of the modern types of vertical axis wind turbines is the Savonius rotor (Fig. 3.18). This consists of a hollow cylinder that is cut along its long axis to form two semi-cylinders. The two halves are mounted into a rough "S" shape so that the wind flows through the cavity, being directed from the back of the concave side onto the inside of the convex side, resulting in rotation.[24]

Savonius wind machines have a low cut-in speed and can operate in winds as low as 5 mph. This makes the machine suitable for electricity generation in low-power applications such as individual domestic installations. The machine is particularly suited to locations of variable wind direction. A Swiss company markets a 6-kW version of the Savonius machine. The peak efficiency of this form of turbine is about 30% (see Fig. 2.5) and the tip-speed ratio is low.

A disadvantage of the Savonius design is its high-solidity factor. Also, the machine is heavy if metal vanes are used. Because of the nature of the

62 *Electricity Generation Using Wind Power*

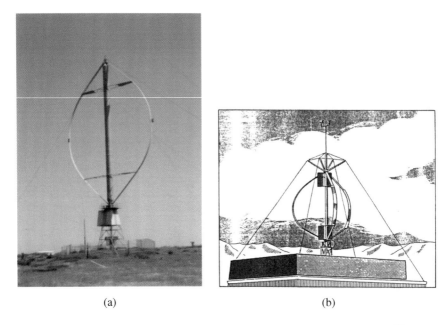

Fig. 3.19 Darrieus wind turbine (a) two-blade rotor (b) three-blade rotor.[20]

construction, the vane or sail area cannot be modified, so that the machine may need to be tied stationary in high winds.

3.6.2 *The Darrieus design*

Much attention and research effort has been devoted to the Darrieus wind machine. This looks like an egg-beater or food-mixer and consists of two or three vertically mounted vanes with aerofoil cross-sections. The shape of the vanes (Fig. 3.19) is the natural shape that a flexible cable, such as a skipping rope, would adopt if it was swung horizontally about the bottom and top pivots.

Unlike the Savonius rotor, the Darrieus machine is not self-starting. The fixed-pitch rotor must be in rotation before the wind exerts a driving force on it. In practical designs, a Savonius rotor is often incorporated onto the Darrieus shaft to provide a starting torque. At high speeds, a Darrieus machine produces far more power than a Savonius machine and has a much higher tip-speed ratio (see Fig 2.7).

Darrieus machines have been studied for single-dwelling domestic housing applications. A blade diameter of about 15 ft is needed to produce 1 kW of output power. The Darrieus design has a peak efficiency of about 35%. A government-sponsored and large-scale Darrieus turbine was developed at the Sandia National Laboratories, Albuquerque, New Mexico, the USA in the early 1980s. This used a blade diameter of 55 ft to develop a power of 80 kW. Another three-blade Darrieus machine in the USA had developed 500 kW. The world's largest Darrieus project, in Quebec, Canada, has developed a 4-MW machine. This is comparable in rating with the largest horizontal axis propeller systems.

As an alternative to Savonius assisted start-up, a Darrieus system can be started by using its coupled generator as a starter motor, taking electrical power from the battery or electricity supply.

Darrieus machines have a low-solidity factor, comparable with that of a horizontal axis propeller system, which makes for an economical use of aerofoil materials. The centrifugal forces of rotation exert tensile stresses on the blades, which may be metallic or made of fibreglass. The forces on the blades are similar in some respects to the aerodynamic forces on an aeroplane wing section or aerofoil.

It should be noted that the detailed operation of both the Savonius and Darrieus design is very complex. Analyses cannot be undertaken using the comparatively simple equations of Secs. 2.2 and 2.3 of Chap. 2.

3.6.3 *Other forms of vertical axis machine*

There are many different forms of modern vertical axis machine. The most promising of these are derivations of the Darrieus principle. For example, Musgrove has designed machines with rotors of a two-blade or three-blade "H" configuration. Two vertical blades can either use a fixed tilt angle with rigid fixing to the hub or be hinged for variable tilt angles. At high speeds, the variable tilt blades move outwards due to centrifugal action and act to govern the speed of rotation, eliminating the danger of over-speeding.[25]

References

1. Eldridge, F. R., *Wind Machines*, Second Edition, Van Nostrand Reinhold Company, New York, the USA, 1980.

2. Dorf, R. C., *Energy Resources and Policy*, Addison-Wesley Publishing Company Inc., New York, the USA, 1978.
3. 'Illustrated History of Wind Power Development', Part 2 — 20th Century Developments, 1996–2002 TelosNet Web Development, available at http://telosnet.com/wind/20th.html.
4. Putnam, P. C., *Power from the Wind*, Van Nostrand, New York, the USA, 1948.
5. Golding, E. W., *The Generation of Electricity by Wind Power*, E and F Spon Ltd., London, England, 1955.
6. Gipe, P., *Wind Energy Basics*, Chelsea Green Publishing Company, White River Junction, Vermont, the USA, 1999.
7. Manwell, J. F., J. G. McGowan, A. L. Rogers, *Wind Energy Explained*, Jhon Wiley & Sons Ltd., Chichester, England, 2002.
8. Reed, J. W., R. C. Maydew, B. F. Blackwell, Wind Energy Potential in New Mexico, SAND-74-0077, Sandia National Labs, Energy Report, New Mexico, the USA, 1974.
9. World Financial Markets, Financial Times Business Information Ltd., London, England, 1983 and 1984.
10. *Renewable Energy World*, **4**(3), 2001.
11. Brown, L. R., Europe Leading World into Age of Wind Energy, Earth Policy Institute, USA, April 2004, available at http://www.earth-policy.org/updates 37.htm.
12. IEE News, Institution of Electrical Engineers, London, England, 1994.
13. *Renewable Energy World*, **7**(2), p. 23, 2004.
14. *Renewable Energy World*, **7**(1), p. 105, 2004.
15. *Renewable Energy World*, **7**(3), 2004.
16. Shepherd, W., D. W. Shepherd, *Energy Studies*, Second Edition, World Scientific Publishing Co. Pte. Ltd.
17. Future Energies, Part of Fuel Cell Network (UK), March 2004, available at http:www.futureenergies.com/article.php?sid.
18. *Renewable Energy*, **7**(5), p. 141, 2004.
19. Feasibility of Large Wind Turbines in the UK, Electrical Research Association (ERA), Leatherhead, Surrey, England, 1974.
20. Carbon Commentary: Trends in UK domestic electricity use, available at www.carboncommentary.com/2008/02/20/76.
21. UK Energy in Brief 2008, Department of Trade and Industry, London, UK, April 2009.
22. Burton, T., D. Sharpe, N. Jenkins, E. Bossanyi, *Wind Energy Handbook*, John Wiley & Sons Ltd., Chichester, England, 2001.
23. Freris, L. L., *Wind Energy Conversion Systems*, Prentice Hall International (UK) Ltd., Hertfordshire, England, 1990.
24. Inglis, D. R., *Windpower and Other Energy Options*, University of Michigan Press, the USA, 1978.
25. Musgrove, P. J., I. D. Mays, The Variable Geometry Vertical Axis Windmill, Proceedings 2nd International Symposium on Wind Energy Systems, Cranfield, England, 1978.

CHAPTER 4

The Location and Siting of Wind Turbines

4.1 The Availability of Wind Supply

In most geographical locations there is a characteristic pattern of wind velocity over the year. The detailed features of the wind-energy resource are well documented in Europe and in North America. There are typical patterns of wind flow on a seasonal basis. At any given site there may be significant variations of wind on a daily (sometimes hourly) basis, with regard to both the magnitude and the direction of the wind, but the annual features are fairly consistent.

4.1.1 *Global survey*

A very broad representation of the average overall global wind is shown in Fig. 4.1, in terms of wind power density in watts/square metre. In the Northern Hemisphere, the wind power density is of order 2–3 times greater in the winter than in the summer. An alternative much quoted picture is shown in terms of shading in Fig. 4.2. In Europe the prevailing wind (i.e., the direction of the wind for most of the time) is westerly. It flows from west to east. The wind map shows that the western coastlines of North America, South America, Northern Europe, and Africa have great potential for the

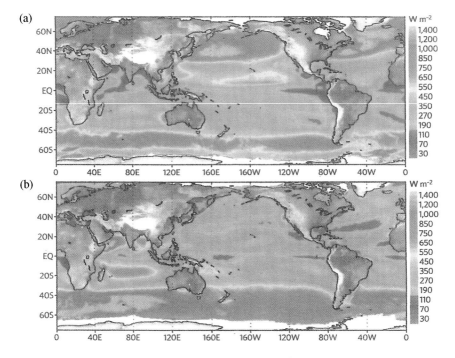

Fig. 4.1 Average power density of the wind (A) Boreal winter (B) Boreal summer [quoted from "Electricity without carbon" Nature, Vol 454, 14 August 2008].

exploitation of wind energy. The annual average wind at any site depends on:

(a) geographical position,
(b) the detailed location and, in particular, the altitude and the distance from the sea,
(c) the exposure and possible screening due to any surrounding hills, vegetation and buildings, and
(d) the shape of the land surface in the immediate area.

In order to gain altitude and thereby increased wind speed, turbine sites are often chosen on hill tops. This is discussed in more detail in Sec. 4.3.

4.1.2 *Energy content of the wind*

A rough estimation of the wind-energy capability of any particular site is provided by the wind speed–duration data. The average wind speed in mph

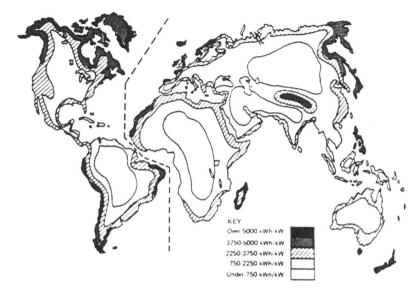

Fig. 4.2 Annual availability of wind energy in different parts of the world.[1] In terms of estimated number of kWh/Year per rated kilowatt output for wind machines designed for rated wing speeds of 25 miles per hour.

is plotted against the number of hour's duration for the $365 \times 24 = 8,760$ hours of the year (Fig. 4.3).[1] It should be noted that a characteristic of power versus time means that the area under the curve has the dimension of power multiplied by time, which is energy. For wind-energy use, the power versus time availability characteristic should contain the largest possible area.

The cut-in speed, which is shown in Fig. 2.10 of Chap. 2, eliminates the hours representing too-low a wind speed. Correspondingly, there are periods where the wind speed is too high and the turbine operation has to be discounted. The hours representing this condition are shown to the left of the "furling point" in Fig. 4.3.

A measure of the perception of wind of different speeds is provided by the Beaufort Scale, shown in Table 4.1. For example, a wind speed of 25 mph is described as a "strong breeze", in which it would be difficult to manipulate an umbrella. The original scale was proposed by Admiral Sir Francis Beaufort of the British navy in 1805. The terminology of the categories of the higher wind speeds has more recently been slightly changed. A category of "hurricane" has been added for wind speeds of 73+ mph.

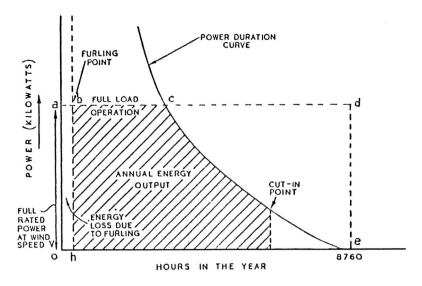

Fig. 4.3 Estimation of annual energy output.

Typical wind velocity–duration characteristics are shown in Fig. 4.4 for a wide range of geographical locations, based on the specified values of average annual wind speed. A more detailed picture, giving month-by-month figures, is given in Table 4.2. For most locations, even in the Southern Hemisphere, the wind speed in the months October–March is significantly greater than in the periods April–September. Since the energy potential and the power content of the wind are proportional to the cube of the wind speed, there will obviously be great differences in the energy viability of the different locations.

4.1.3 Wind-energy supply in Europe

A European wind atlas of comparative wind speeds is shown in Fig. 4.5. The wind-energy resources vary widely over continental Europe and are influenced by three major factors:

1. Large temperature differences between the polar air in the north and sub-tropical air in the south.
2. The distribution of land and sea with the Atlantic Ocean to the west, Asia to the East, and the Mediterranean Sea and Africa to the South.

Table 4.1 Beaufort wind force scale (Ref. http://www.bunganutlate.org/beaufortscale.htm).

Beaufort scale	Description	When You See This	Wind Speed mph	Wind Speed kmh
0	Calm	Smoke goes straight up. No wind present	Less than 1	Less than 2
1	Light Air	Direction of the wind is shown by smoke drift, but not by a wind vane.	1–3	2–5
2	Light Breeze	Wind is felt on your face. Leaves rustle. Wind vane moves.	4–7	6–11
3	Gentle Breeze	Small twigs and leaves move steadily. Wind extends small flag straight out.	8–12	12–19
4	Moderate Breeze	The wind raises loose paper and dust. Samll tree branches move.	13–18	20–29
5	Fresh Breeze	Waves form on lakes and ponds. Small trees sway.	19–24	30–39
6	Strong Breeze	Arge tree branches move. Umbrellas become hard to use. Wires whistle.	25–31	40–50
7	Moderate Gale	Hard to walk against the wind. Whole trees are in motion.	32–38	51–61
8	Fresh Gale	Very difficult to walk against the wind. Twigs break from trees.	39–46	62–74
9	Strong Gale	Roof shingles are torn away. Small damage to buildings occur.	47–54	75–87
10	Wholw Gale	Trees are uprooted.	55–63	88–101
11	Violent storm	widespread damage from the wind.	64–72	102–116
12	Hurricane	Widespread destruction from the wind.	73+	117+

3. The major geographical features such as the Alps, the Pyrenees, and the mountains of Scandinavia.

The British Isles consisting of Great Britain (England, Scotland, Northern Ireland, and Wales) and the Republic of Ireland form one of the windiest regions on the earth. Contours of the mean annual wind speed, measured

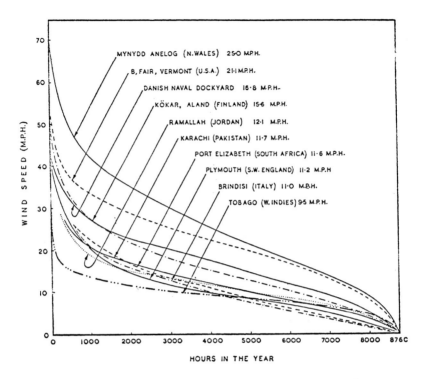

Fig. 4.4 Velocity-duration curves for widely separated sites.[1]

at the agreed standard height of 33 ft (10 m) above the ground, are given in Fig. 4.6.[1] The wind speed increases greatly at higher levels of elevation.

In general, the coastal areas are windier than inland areas. The prevailing wind is westerly, from the Atlantic Ocean, creating the high average value of 17.5 mph along the western coastlines of Scotland and Ireland. Incidentally, the same regions offer a great potential for water-wave energy.

Tremendous local variations of wind energy occur even in a relatively small country such as Great Britain. In Fig. 4.7, the units of wind speed are plotted against the hours of wind availability per year for three different UK locations. It should be noted that curve A in Fig. 4.7 represents a location on the west coast of Scotland where the wind is almost identical to that for the top curve in Fig. 4.4, representing a location in N. Wales. Characteristics of wind power versus time duration for the same three UK locations are shown in Fig. 4.8. Obviously, the preferred location is Rhossili Down, which is on

The Location and Siting of Wind Turbines 71

Table 4.2[1]

Place of region	Monthly mean wind speed in m.p.h												Yearly mean speed (m.p.h.)	Percentage departures from the annual mean		No. of years of observations	Height of anem. above ground (feet)
	Jan.	Feb.	Mar.	Apr.	May	June	July	Aug.	Sep.	Oct.	Nov.	Dec.		Maximum	Minimum		
Northern hemisphere																	
England and Wales*	14.4	13.8	12.5	12.3	10.8	10.7	10.6	9.9	10.8	12.8	13.1	13.5	12.1	19.0	18.2	15	33
Paris (Eiffel Tower)†	22.8	22.0	20.8	19.6	18.7	17.0	16.6	17.9	18.4	20.8	20.7	22.0	19.8	15.2	16.2	Long period	1,000
Moscow	10.8	10.6	10.3	9.6	9.7	8.9	8.6	9.1	9.9	10.3	10.1		9.7	11.3	11.3	10	—
Berlin	13.2	12.3	12.1	11.4	10.8	10.1	9.6	10.3	9.9	10.5	11.7	12.3	11.3	16.8	15.0	30	30
Prague	6.7	6.2	5.7	10.0	7.2	7.4	6.5	6.4	7.2	6.4	6.2	5.1	6.7	49.3	23.9	—	—
Brindisi	10.8	12.2	13.5	9.8	9.6	10.2	11.1	11.1	9.8	10.4	11.2	12.6	11.0	22.7	12.7	3	Surface
Tripoli	11.2	9.4	9.4	10.3	10.2	9.2	8.0	8.9	9.5	8.4	10.6	10.5	9.6	16.7	16.7	3	Surface
Ramallah (Jordan)	16.2	13.2	15.3	12.6	11.3	11.9	12.6	10.7	10.0	9.5	10.0	11.3	12.1	36.0	21.5	4	70
Seistan	4.2	5.2	5.7	7.5	9.6	11.7	14.2	13.7	11.9	6.9	5.2	4.0	8.3	71.0	51.8	Long period	22
Perim (Adan)	14.5	14.0	14.1	13.1	10.0	9.8	11.2	10.1	8.8	13.5	14.7	14.1	12.3	19.5	28.5	Long period	—
Karachi	7.7	8.8	11.3	12.2	14.5	16.6	16.4	14.9	11.7	8.2	6.6	11.4	11.7	42.0	43.6	Long period	44
U.S.A. (New England)	11.0	11.2	11.4	11.0	9.8	9.0	8.6	8.4	8.8	9.6	10.5	10.6	10.0	14.0	16.0	Long period	22 stations

(*Continued*)

72 Electricity Generation Using Wind Power

Table 4.2 (Continued)

Place of region	Monthly mean wind speed in m.p.h												Yearly mean speed (m.p.h.)	Percentage departures from the annual mean		No. of years of observations	Height of anem. above ground (feet)
	Jan.	Feb.	Mar.	Apr.	May	June	July	Aug.	Sep.	Oct.	Nov.	Dec.		Maximum	Minimum		
Southern hemisphere																	
Wellington (N.Z.)	9.3	9.7	8.9	8.1	8.0	8.2	7.5	8.2	9.3	10.2	10.1	9.5	8.9	14.6	15.7	3	35
Port Elizabeth (S.A.)	12.2	11.8	11.5	10.8	9.6	10.0	10.2	11.0	12.6	13.3	13.3	12.9	11.6	14.6	17.2	8	50
Perth (W. Australia)	13.8	13.5	12.8	10.7	10.6	10.6	11.2	11.8	11.8	12.6	13.4	13.9	12.2	13.9	13.1	20	71
Sydney (N.S.W.)	8.9	8.1	7.5	7.0	6.8	7.1	7.2	7.4	8.0	8.2	8.5	8.9	7.8	14.1	12.8	26	58
Adelaide (S. Australia)	9.9	8.8	8.3	8.0	8.1	8.3	8.5	9.2	9.2	9.8	9.9	9.9	9.0	10	11.1	30	75
Ascuncion (Paraguay)	2.9	3.0	3.0	3.2	3.6	4.0	4.6	4.7	5.0	4.5	3.5	3.0	3.7	35	22	—	—
Rio Gallegos (Argentina)	9.3	10.5	9.9	9.5	7.8	7.3	5.9	5.8	8.4	10.1	11.7	10.5	8.8	33	34	—	—
Buenos Aires	10.1	10.5	8.9	9.9	8.4	8.8	9.8	10.0	11.0	10.1	10.3	10.2	9.8	12	14	—	—
Campos (Rio de jan.) Brazil	9.2	10.3	7.4	5.6	6.0	6.9	6.3	7.4	9.2	7.4	9.6	8.5	7.8	32	28	—	—
Recife (Pernambuco) Brazil	13.6	13.0	11.6	12.5	15.9	14.5	13.6	13.6	14.5	13.6	13.0	12.5	13.6	17	15	—	—

* Average for 20 stations.
† At 1,000 ft.

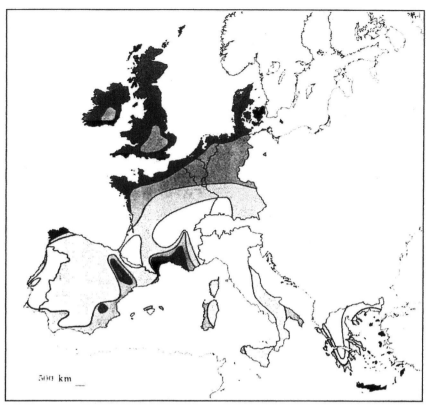

Fig. 4.5 European wind resource map. (Source: Risø Laboratory Denmark.)

the west coast of Scotland, whereas a location in the English Midlands such as Leicester is completely unsuitable.

There are important seasonal variations in the availability of wind energy in the UK. In northern temperate latitudes, the highest daily average wind speeds occur in winter, with maximum values usually being in January, and the lowest in July/August. The seasonal variation between winter and summer is of the order 2:1. The average wind-energy flux density on a typical

74 Electricity Generation Using Wind Power

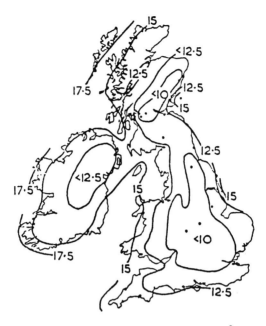

Fig. 4.6 Isovent contours of UK mean wind speed (mph)² (courtesy of the UK Meteorological Office).

month-by-month basis for the UK is given in Fig. 4.9.[2] It is of immense significance in the UK that the availability of wind energy closely matches the national demand for electricity. It is also relevant that the profile of UK wind-energy input is in anti-phase with the profile of solar radiation input.

4.1.4 *Wind-energy supply in the USA*

The mean annual wind speeds across the USA are shown in Fig. 4.10. The predominant feature is the southerly winds blowing north through Texas to North Dakota along the entire N–S dimension (about 1,500 miles). This is reflected in the wind-power density data of Fig. 4.11, which shows that the centre and western states of the USA have greater wind-energy potential than the eastern states. Many of the eastern states have no significant wind-energy potential at all. In Fig. 4.12, the electrical-energy potential of the wind is shown as a percentage of the total US electricity consumption in 1990.[3] These figures suggest that the central states of (from S to N) Texas,

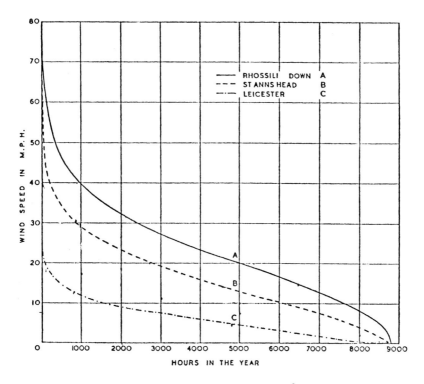

Fig. 4.7 Velocity duration curves.[2]

Oklahama, Kansas, Nebraska, and the Dakotas have the potential to supply almost all of the US demand of electricity by wind energy alone. The potential is obviously such that the USA is wise to consider wind energy as a major renewable resource.

Typical figures of the annual average wind speeds in the four major cities Cincinnati, Columbus, Cleveland, and Toledo of the US state of Ohio, which lies immediately south of the Great Lakes are shown in Table 4.3. In Fig. 4.10, the mean annual wind-power density for Ohio is shown as $>200\,W/m^2$. The southernmost large city Cincinnati, in the SW corner of the state, has a mean wind speed of 7.1 mph, which is too low for consideration as a site for wind-energy installations. The northernmost city of Ohio is Cleveland on the southern shore of Lake Erie. This has a mean annual wind speed of 10.9 mph and is therefore a very viable site as a location for raising wind-powered electricity. The information of Table 4.3 is taken from the

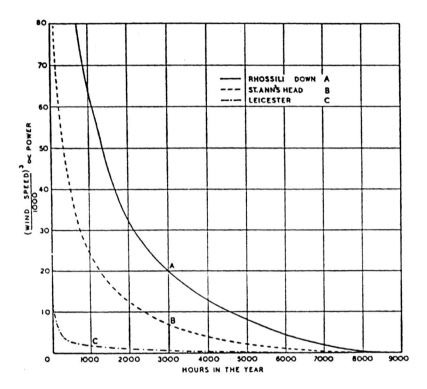

Fig. 4.8 Wind power-duration characteristics for three UK locations.[2]

very reliable data issued annually by the US Meteorological Office. It shows monthly average wind speeds for all the months of the year from January to December in addition to the mean annual value. For all of the cities listed, the wind speed is lowest in the summer months and significantly higher in the winter months October to May. The monthly profile of wind-flux density will therefore follow the UK pattern shown in Fig. 4.8. In Ohio, the wind-energy flux pattern correlates well with the state demand for electricity.

Also shown in Table 4.3 is the wind data for Chicago, which lies at the SW corner of Lake Michigan. Although Chicago is known colloquially as the "windy city", the data shows that Cleveland is significantly windier in most months of the year. The ratio of mean annual wind speeds between Cleveland and Chicago is seen to be 10.9/10.4 which is 1.048. But since the wind power is proportional to the cube of the wind speed, the ratio of wind

The Location and Siting of Wind Turbines 77

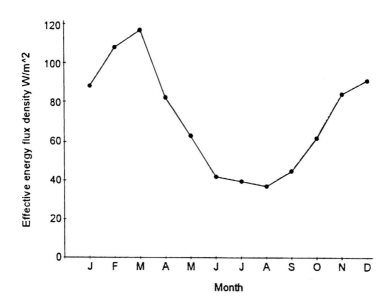

Fig. 4.9 Wind availability in the UK (mean wind velocity = 6 m/s = 13.4 mph).[4]

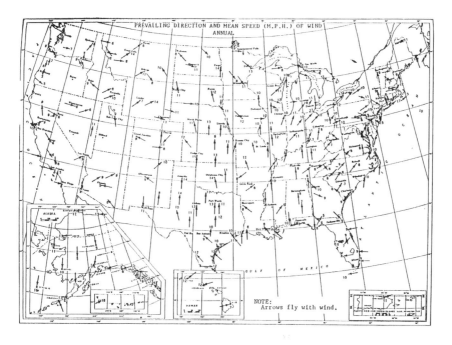

Fig. 4.10 US wind atlas (courtesy of the US Meteorological Office).

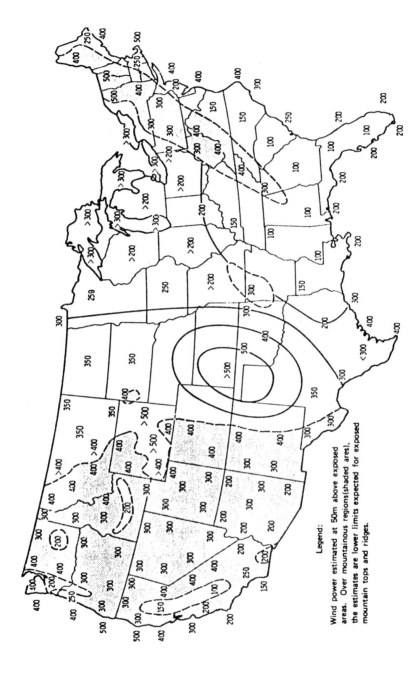

Fig. 4.11 Mean annual wind power density (W/m^2).[1] *Source*: Battelle Pacific Northwest Laboratories.

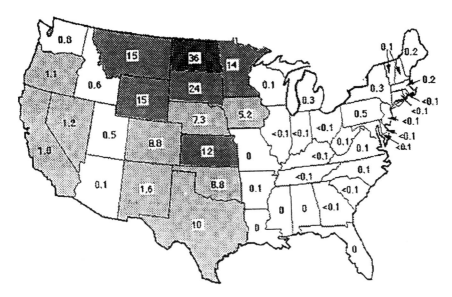

Fig. 4.12 Electrical energy potential of the wind as a percentage of the total US electrical consumption 1990.[5]

power represented is $(10.9/10.4)^3$ that is 1.15. In other words, Cleveland represents a possible power and energy gain of 15%, which is very large!

4.2 Statistical Representation of Wind Speed

Although the year-to-year variation in annual mean speed is not easily predictable, the speed variations at a given site during the year can be characterised statistically in terms of a probability distribution.

A histogram showing the probability of wind speeds over a range of $0 < V < 20$ mph at a certain location is shown in Fig. 4.13.[4] In this example, the wind speed was measured in 1 m/s sections and the sum of the heights of the 20 samples is 100%. If the sample width is made smaller and more samples taken, the curve of the profile becomes smoother. In the limit, with a large number of samples, the profile curve is continuous and is called the probability density function. Figure 4.13 shows the statistical probability of the wind speed V being in a certain 1 m/s interval. For example, the probability of the wind speed being between 6.5 and 7.5 m/s is 12% or 0.12 per-unit, which means $0.12 \times 8,760$ or 1,051.2 hours/year at this site.

Table 4.3

	Cincinnati								
	Relative humidity				Wind Ø				
							Fastest mile		
	Hour	Hour (Local time)	Hour	Hour	Mean speed	Prevailing direction	Speed	Direction	Year
					43	40	50	50	
JAN					8.3	SW	49	SW	1959
FEB					8.4	SW	49	SW	1918
MAR					9.0	SW	49	SW	1922+
APR					8.4	SW	47	SW	1920
MAY					6.7	SW	36	W	1945
JUN					6.4	SW	40	W	1934
JUL					5.2	SW	43	SW	192
AUG					5.1	SW	38	W	1944
SEP					5.4	SW	38	SW	1941
OCT					6.1	SW	35	SW	1937
NOV					7.7	SW	47	SW	1919
DEC					7.9	SW	41	SW	1920
YEAR					7.1	SW	49	SW	JAN. 1959+

	Columbus								
	Relative humidity				Wind				
							Fastest mile		
	Hour 01	Hour 07	Hour 13 (Local time)	Hour 19	Mean speed	Prevailing direction	Speed	Direction	Year
	12	12	12	12	22	14	20	20	
JAN	74	76	67	69	10.1	SSW	56	W	1959
FEB	74	77	65	68	10.3	NW	57	W	1956
MAR	70	74	59	62	10.7	SSW	63	NW	1955
APR	71	76	53	56	10.0	WNW	56	W	1970
MAY	77	79	55	58	8.5	S	54	NW	1964
JUN	80	81	53	58	7.3	SSW	47	NH	1966
JUL	82	84	56	59	6.6	SSW	49	NW	1956+
AUG	85	88	57	63	6.2	NNW	43	NW	1965
SEP	84	88	58	65	6.7	S	38	N	1963
OCT	79	83	55	64	7.6	S	39	SW	1965
NOV	79	82	65	71	9.5	S	61	N	1952
DEC	77	79	69	73	9.7	W	47	SW	1971
YEAR	78	81	59	64	8.6	SSM	63	NW	MAR. 1955

(*Continued*)

Table 4.3 (*Continued*)

	Cleveland								
	Relative humidity				Wind				
	Hour 01	Hour 07	Hour 13	Hour 19	Mean speed	Prevailing direction	Fastest mile		
		(Local time)					Speed	Direction	Year
	11	11	11	11	30	14	30	30	
JAN	73	75	69	71	12.4	SW	68	SW	1959
FEB	75	78	69	71	12.4	S	65	W	1958
MAR	75	78	65	69	12.5	W	74	W	1946
APR	71	74	57	60	11.9	S	65	W	1950
MAY	73	75	56	57	10.5	S	68	SW	1951
JUN	78	78	57	59	9.5	S	57	SW	1954
JUL	80	81	57	61	8.8	S	65	W	1956
AUG	82	84	59	64	8.4	S	45	S	1956
SEP	79	83	59	68	9.1	S	43	W	1951
OCT	76	80	56	67	10.1	S	43	W	1946
NOV	74	77	66	71	12.3	S	59	W	1946
DEC	74	75	70	72	12.4	S	49	SW	1971
YEAR	76	78	62	66	10.9	S	74	W	MAR. 1946

	Toledo								
	Relative humidity				Wind				
	Hour 01	Hour 07	Hour 13	Hour 19	Mean speed	Prevailing direction	Fastest mile		
		(Local time)					Speed	Direction	Year
	16	16	16	16	16	8	16	16	
JAN	72	79	69	73	10.8	WSW	47	W	1971
FEB	72	78	66	70	10.9	WSS	56	SW	1967
MAR	73	81	61	67	10.9	WSW	56	W	1957+
APR	77	80	55	59	10.9	E	72	SW	1956
MAY	76	79	51	56	10.0	WSW	45	W	1957
JUN	82	82	54	58	8.3	SW	50	W	1969
JUL	84	86	55	61	7.5	WSW	54	NW	1970
AUG	86	89	57	65	7.3	SW	47	W	1965
SEP	86	90	57	70	7.8	SSW	47	NW	1969
OCT	81	85	55	68	8.7	WSW	40	SW	1956
NOV	81	84	66	74	10.3	WSW	65	SW	1957
DEC	82	83	73	78	10.5	SW	45	SW	1971+
YEAR	79	83	60	67	9.5	WSW	72	SW	APR. 1956

(*Continued*)

Table 4.3 (Continued)

					Chicago				
		Relative humidity				Wind			
								Fastest mile	
	Hour 00	Hour 06	Hour 12	Hour 18	Mean speed	Prevailing direction	Speed	Direction	Year
			(Local time)						
	8	8	8	8	29	19	29	29	
JAN	69	70	64	66	11.5	W	50	W	1950
FEB	68	70	60	64	11.7	W	51	SW	1967
MAR	68	73	58	60	11.9	W	54	NW	1955
APR	67	72	55	56	11.8	W	50	NW	1951
MAY	67	71	51	51	10.5	SSW	54	S	1950
JUN	68	74	54	54	9.3	SW	50	W	1953
JUL	72	77	55	56	8.3	SW	46	NW	1959
AUG	74	79	55	57	8.1	SW	54	NW	1949
SEP	75	80	55	59	9.0	S	48	SW	1959
OCT	69	77	53	58	9.8	S	45	S	1949
NOV	73	77	63	68	11.4	SSW	60	SW	1952
DEC	75	78	70	72	11.2	W	50	SW	1948
YEAR	70	75	58	60	10.4	W	60	SW	NOV. 1952

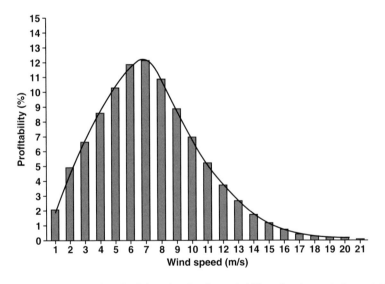

Fig. 4.13 Histogram and Weibull function for the probability of a given wind speed (data measured in 1 m/s units).[6]

For many typical sites, the probability density function representing wind occurrences can be represented, with good accuracy, by the statistical function known as the Weibull distribution $h(V)$, given by:

$$h(V) = \frac{k}{V}\left(\frac{V}{C}\right)^{k-1} \varepsilon^{-\left(\frac{V}{C}\right)^k}. \tag{4.1}$$

Value $h(V)$ in Eq. (4.1) is the frequency of the occurrence of wind speed V, factor C is the characteristic wind speed, sometimes referred to as the scale parameter, and item k is the shape parameter.

The exponential term in Eq. (4.1) represents the probability of the wind speed exceeding value V and is sometimes called the cumulative Weibull Distribution $H(V)$

$$H(V) = \varepsilon^{-\left(\frac{V}{C}\right)^k}. \tag{4.2}$$

Weibull distributions $h(V)$ versus wind speed V for the different values of shape factor k are shown in Fig. 4.14.[5] If the variation of hourly wind speed about the annual mean is small, then the value of k is relatively high, such as $k = 2.5$ or 3. When there is greater variability of wind speed about the mean, then k has the smaller value $k = 1.5$ or 1.25. A comparison of Fig. 4.13 with Fig. 4.14 shows that the value of k in Fig. 4.13 must be close to the value $k = 2$, which is a typical value at many wind sites. For this

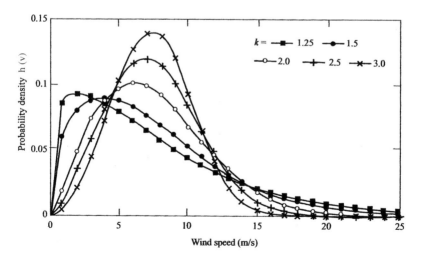

Fig. 4.14 Example Weibull distribution.[7]

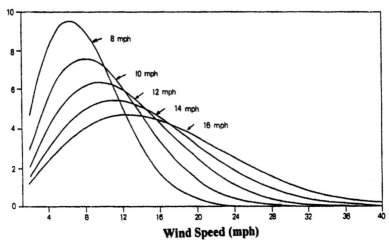

Fig. 4.15 Effect of scale parameter variation C on the Weibull distribution $k = 2$.[8]

value of distribution, more days have lower than the mean speed while few days have high wind speed.

When $k = 2$, the Weibull distribution defined in Eq. (4.1) becomes mathematically equal to the statistical distribution known as the cumulative Rayleigh distribution. The effects of changing the scale parameter C for a fixed value of shape parameter k on the Weibull (and Rayleigh) distributions are shown in Fig. 4.15.[6]

The parameters C and k for the Weibull frequency distribution can be found from Eq. (4.2) by plotting the exponential logarithms $\ln V$ versus $\ln(-\ln H(V))$. The log–log plot fits a good straight line in which the slope of the line is k and value C is equal to the intercept where the characteristic intersects the $\ln V$ axis.[4]

Although a statistical simulation of wind speeds is helpful at an early stage of design, it must be regarded only as a first and approximate step. It is essential to have measured data.

4.3 Choice of Wind Turbine Sites

The wind speed information in Figs. 4.5 and 4.10 gives a good and first-step picture of the areas of a country that might be suitable for wind turbine location. However, in order to select a particular site, onshore or offshore, very much more detailed information is required. It is not possible, in general, to accurately calculate the wind speed at a particular point by theoretical means.

The only way to be sure of the actual wind speed at any fixed location is to measure it. Moreover, it is necessary to take a series of readings over a time span of a year or more in order to accommodate seasonal changes. Any theoretical values of long-term wind speed determined by modelling, statistical methods, wind atlas calculations, or other theoretical means must be verified by on-site measurements.

The siting of an individual turbine or a group of turbines, for electricity utility interconnection, involves the various issues described below.[7]

4.3.1 Identification of suitable areas

Various methods such as a wind atlas are available to determine the geographical areas that have high average wind speed at the proposed hub height of the turbine. The known minimum assured wind speed must be appropriate for the types and designs of turbine proposed.

4.3.2 Selection of possible sites within the chosen area

Sites must be adequate from the viewpoint of engineering installation, public acceptability, and legal compliance. Access to a site is required for heavy-duty vehicles by road, rail track, or sea (for offshore installations). Any site must be characterised by appropriate land slopes and contours at the turbine site, as discussed in Sec. 4.4. Issues of land ownership, zoning laws, and planning permissions need to be assured. Certain geological matters arise such as foundation design and stability, the ground electrical resistance for lightning protection, and the possibility of site erosion. Any proposed site is likely to be subjected to environmental scrutiny from the viewpoints of protection of wildlife and also of the visual and audible impact of the turbines. There has to be complete independence from microwave communication systems. The electrical power lines of suitable voltage and current ratings, with their control and protection systems, must be accessible of interconnection. The turbine site must be at an acceptable distance from residential

Table 4.4 Features affecting the choice of a wind turbine site.

Wind resource issues:

— Average wind speed at the hub height
— Seasonal wind variations
— Wind power-hours profile
— Incidence and size of wind gusting and the possibility of windstorms
— Contours of the site terrain, for aerodynamic effects
— Clustering effects, for multi turbine sites
— Exposure to the prevailing wind, screening effects of surrounding hills, vegetation, buildings

Legal issues:

— Leasing and planning permissions
— Ownership of the site and surrounding area
— Location of access facilities
— Noise abatement legislation

Environmental issues:

— Proximity to residential areas
— Proximity to places of historical, archaeological, or cultural interest
— Proximity to wildlife sanctuaries, breeding grounds or migration routes
— Can the site be concurrently used for crop farming or animal grazing?
— Acoustic noise of the gearbox and rotor blades
— Visual impact on the scenery
— Electromagnetic interference with telecommunication systems (e.g., caused, by rotation of the rotor blades)

Access issues:

— Road, rail track, and water access capable transporting heavy loads of?
— Construction costs of the towers and access roads
— Is it necessary to guard the installation against randalism or terrorism?

Engineering issues:

— Adequacy and structural stability of the site foundation
— Proximity to the electrical transmission system with which it is to be interconnected
— Ground electrical resistance for lightning protection

areas and from the places of historical, archaeological, and cultural importance.

The features to be considered in a site selection are listed in Table 4.4. These features are not listed in priority order but each is sufficiently important to justify consideration as an individual item.

4.4 Effects of the Site Terrain

The aim in turbine location is always to maximise the velocity of a smooth and controlled flow of air to the turbine. Turbine sites are usually on high ground and often on hilltops. A hilltop of suitable shape can be used to enhance the local wind speed in a similar manner to the design of an aerofoil in a laboratory wind tunnel. The streamlines of an air stream are compressed to represent the acceleration of the airflow, as it passes over a smooth hilltop profile (Fig. 4.16). A hilltop terrain with abrupt sides would cause turbulence and might involve flurries of reverse flowing air on the hilltop, which would be quite unsuitable for driving a turbine. Similarly, hills with sharp peaks might have large upstream wind velocity that would not be sustained at the turbine location on the hilltop.

It is a well-known experience that the wind-flow velocity may be accelerated in a valley or a canyon, depending on the topology of the surrounding hills. Some ancient forms of vertical axis wind machine were built in culverts to engage this advantage. In city centres with a lot of high-rise buildings, not necessarily skyscrapers, wind funneling occurs in the streets dividing the buildings and can be a hazard to pedestrians during high-wind conditions.

By appropriate site location, the funneling effect of a culvert or canyon can be used to increase the wind velocity at the rotor of a wind turbine. In some cases, excavation and contouring of the ground may be used to enhance the local wind speed, as illustrated in Fig. 4.17.[8]

A combination of the culvert effect incorporating aerofoil designed walls, with a diffuser tailpiece, is shown in Fig. 4.18. This increases the amount of wind energy received at the turbine. Such a design can be used to decrease the turbine diameter for a given power rating and also to increase

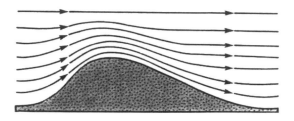

Fig. 4.16   Acceleration of wind over fill.[1]

88 *Electricity Generation Using Wind Power*

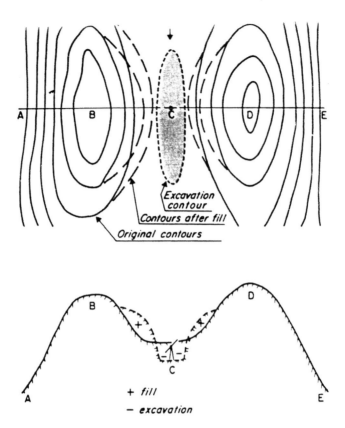

Fig. 4.17 Proposed type of terrain modification for the purpose of augmenting average wind speeds.[8]

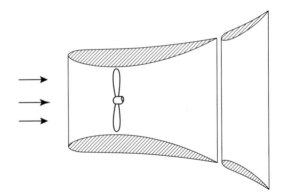

Fig. 4.18 Ducted turbine with diffuser.[1]

Fig. 4.19 Vortec offshore turbines.[4]

the rotational speed, thereby reducing the required step-up gear ratio to the generator.

The culvert effect has been designed into a range of offshore wind turbines shown in Fig. 4.19. In this design, the turbines have a conical-shaped rotor housing, referred to as a "diffuser augmentor". Individual turbines are each rated at 5 MW and have a rotor diameter of m.[9]

4.5 Spacing Effects of Wind Farm Arrays[10]

A group of wind turbines on the same site is usually called a "wind farm" or a "wind array". The individual turbines are usually linked electrically and commercially. From the viewpoint of repair and maintenance, there are obvious advantages in having turbines in close proximity on the same site. In addition, a wind farm or a group of several turbines on the same site can enjoy certain electrical engineering advantages, which are discussed in Chaps. 8 and 9.

The number and spacing of turbines on a site are major design issues. Sometimes, it is not possible for all the turbines to be in ideal free-wind locations, accessing the uninterrupted flow of the wind stream. In addition, the wind flow may vary across a wind farm site due to differences of terrain and/or differences of upstream obstructions. As described in Chap. 2, the action of a turbine is such that the wind velocity and energy downstream are reduced compared with the upstream values. Also, the downstream machines may encounter increased turbulence due to the actions of turbines further upstream. This decreases the energy production of the downstream machines and can also promote the onset of fatigue in the rotors. Wind turbine spacing on a site affects fluctuations of the power output, which can, in turn, affect the electrical network to which it is connected.

The total energy output from an array is lower than would be produced by summing the outputs from each individual turbine if it was located in an optimal upstream location. Array energy losses depend on the characteristics of the local wind, the wind turbine operating characteristics, the number of turbines, and the area of the array. But the most important feature affecting wind-farm losses is the turbine spacing, both downwind and crosswind, illustrated in Fig. 4.20.

There is no scientific and systematic analytical procedure for calculating the spacing of the turbines on a site. This multivariable problem is not amenable to neat formula solution. Design has to be approached on an empirical basis, based on the practical experience and on the various

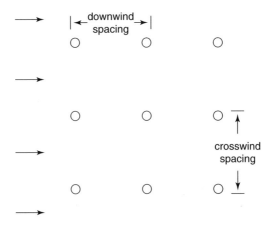

Fig. 4.20 Schematic wind farm array.

theoretical model solutions available in the technical literature. Some of these are described in Ref. 7.

One study indicates that, for turbines spaced 8 to 10 rotor diameters apart in the downwind direction and 5 rotor diameters apart in the crosswind direction, array losses are typically less than 10%.[7] Although this energy loss is only small in percentage terms, it probably represents a large amount of energy and can be a critical factor in the economics of the site operation.

A good estimate of the land (or sea) area covered by a wind farm of large turbines is to use the value 200 ft as a typical rotor diameter for a high-power wind turbine. In the downwind direction, the spacing required for ideal clearance from the neighbouring upwind turbine effects is the value 8–10 times 200 or 1,600–2,000 ft. For a wind array of three ranks downwind, as shown in Fig. 4.20, the downwind distance is about 4,000 ft or about three quarters of a mile. The 3×3 array of Fig. 4.20 therefore covers a surface area of $4,000 \times 4,000 \, \text{ft}^2$ which is roughly one-half of a square mile. In agricultural terms, the surface area represents about 367 acres of land. Where land is scarce and expensive, as in the UK, the purchase, appropriation and the use of the site are obviously important parts of wind-farm operation and financing.

The 140 turbines of the Whitelee wind farm, nine miles south of Glasgow, Scotland, that is briefly discussed in Chap. 3, occupy a windy site of 5,500 hectares or 13,600 acres. Most of this moorland is usable for farming or recreational purposes right up to the base of the towers.

4.6 Problems and Review Questions

4.1 List and briefly discuss the factors that you would take into consideration in selecting a site for a land-based wind machine.

4.2 What order of average wind speed would you expect to find in a "windy" location within the UK? Sketch the form of the annual variation (January to December) of the effective wind-energy flux density in the UK and comment on this.

4.3 What regions of the USA appear to be the most promising for the further development of wind-power generation?

4.4 The US city of Chicago, at the southern end of Lake Michigan, has an average wind speed of 10.4 mph and is known as the "windy city". Use figures from the US Meteorological Office in Table 4.3 to compare the Chicago data with corresponding figures for Cleveland, Ohio, which is on the southern shore of Lake Erie.

4.5 Tabulate and discuss the arguments used in favour of and in opposition to large-scale land-based wind turbines.

4.6 Why are the economic prospects of wind-generated electricity in the UK partially dependent on the retail price of gasoline (petrol) in the USA?

4.7 What property of the wind is describable by a Weibull distribution?

4.8 Explain the meaning of terms shape factor k and scale factor C for the simulation of wind distribution described by the Weibull distribution.

4.9 Obtain an expression for the Weibull probability density function $h(V)$, when the scale parameter $k = 2$, in terms of wind velocity V and shape parameter C.

4.10 What area of the ground would be needed to build a land-based wind farm of 16 large turbines if they were configured into a 4×4 array?

References

1. Golding, E. W., *The Generation of Electricity by Wind Power*, E. and F. N. Spon Ltd., London, England, 1955.
2. Shepherd, W., D. W. Shepherd, *Energy Studies*, Second Edition, World Scientific Publishing Co. Pte. Ltd., London, England, 2003.
3. Elliott, D. L., M. N. Schwarts, Wind Energy Potential in the United States, National Wind Technology Center, Washington, DC, the USA, 1993, available at http://www.nrel.gov/wind/potential.html.
4. Walker, J. F., N. Jenkins, *Wind Energy Technology*, John Wiley & Sons, UNESCO Energy Engineering Series, Chichester, England, 1997.
5. Burton, T., D. Sharpe, N. Jenkins, E. Bossanyi, *Wind Energy Handbook*, John Wiley & Sons Ltd., Chichester, England, 2001.
6. Patel, M. R., *Wind and Solar Power Systems*, CRC Press, Florida, the USA, 1999.
7. Manwell, J. F., J. G. McGowan, A. L. Rogers, *Wind Energy Explained: Theory, Design and Application*, John Wiley & Sons Ltd., Chichester, England, 2002.
8. F. R. Eldridge, *Wind Machines*, Second Edition, Van Nostrand Reinhold Company, New York, the USA, 1980.
9. *Renewable Energy World*, London, England, **4**(3), 2001.
10. Lissaman, P. B. S., A. Zaday, G. W. Gyatt, Critical Issues in the Design and Assessment of Wind Turbine Arrays, Proceedings 4th International Symposium on Wind Energy Systems, Stockholm, Sweden, 1982.
11. 'European Wind Energy Atlas', Soren Krohn, August 2000, available at http://www.windpower.org/tour/wres/euromap.htm.

CHAPTER 5

Power Flow in Electrical Transmission and Distribution Systems

5.1 Basic Forms of Power Transmission Networks

Electrical power is invariably generated and transmitted to loads using three-phase generators and transmission systems. The choice of three-phase systems with sinusoidal time-varying waveforms is entirely economic. This is the most electrically efficient and the least wasteful system.

The basic components of an electrical power system are usually depicted in the form of a "line diagram", which shows the generators, transformers, transmission lines, and loads. For example, Fig. 5.1 shows a single-generator feeding a range of loads via a high-voltage transmission line. The generator voltage is increased to transmission level being 132 kV and above in the UK by a step-up transformer T. At the receiving (load) end of the transmission, voltage level is stepped down to a suitable distribution level.

Figure 5.2 shows a two-generator system feeding a range of loads, including a synchronous motor (SM) that can be excited to operate as a synchronous capacitor if required, as described in Chap. 6. A more complicated multi-generator system is shown in Fig. 5.3, quoted from Ref. 1.

94 Electricity Generation Using Wind Power

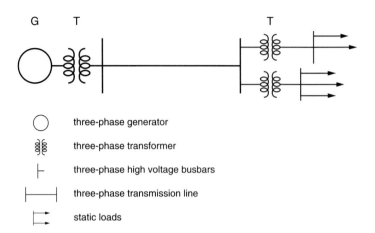

Fig. 5.1 Line diagram of a single generator transmission system.

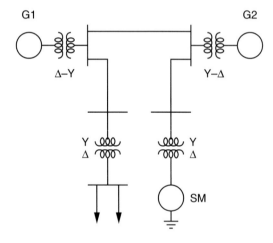

Fig. 5.2 Line diagram of a two-generator system feeding passive loads and a synchronous motor (SM).

The distribution voltage level referred to in this figure is 33 kV, 11 kV, or 6.6 kV in Britain, feeding consumers operating at the domestic levels of 415-V three-phase or 240-V single-phase.

For a detailed exposition of the operation of electrical power systems, the reader is referred to the many existing texts.[1–3]

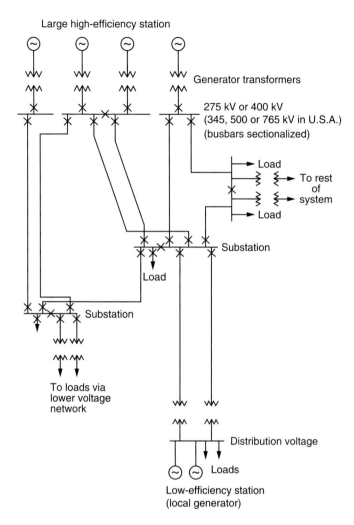

Fig. 5.3 Part of a typical power system.[1]

5.2 Current and Voltage Relationships[4]

When the load on an electricity generation or distribution system is mainly due to lighting or heating systems, the load can be represented in equivalent circuit terms by a resistance. Most industrial loads are partly due to lighting and/or heating devices and partly due to AC electric motors, mainly three-phase induction motors.

Electric motors are inductive in nature so that their equivalent circuit impedances usually contain series inductances. Electricity distribution systems contain transmission lines and transformers, both of which are partly inductive in nature. The result is that the equivalent circuits representing electrical transmission and distribution systems and their loads contain mainly series resistances and inductances.

Modern electricity supply systems are three phase in nature but incorporate the facility to supply both three-phase and single-phase loads. Three-phase loads can usually be evaluated in equivalent single-phase terms. The most general form of representation is a single-phase circuit energised by a sinusoidal voltage source (Fig. 5.4) having the instantaneous value.

$$e(\omega t) = E_m \sin \omega t. \tag{5.1}$$

When the load impedances are "linear", they do not vary significantly in value due to temperature or to current level. The input current $i(\omega t)$ is then also sinusoidal and of the same angular frequency ω as the voltage. If the input impedance to the load has a value Z at the frequency ω, then the current can be written:

$$i(\omega t) = I_m \sin(\omega t \pm \varphi) = \frac{E_m}{|Z|} \sin(\omega t \pm \varphi). \tag{5.2}$$

Any load network can be reduced to a series combination of resistance R and reactance X. The great majority of power system loads is partly inductive and can be represented by a series resistance–inductance (R–L) circuit, as shown in Fig. 5.5. The load impedance magnitude $|Z|$ and time-phase angle ϕ are given by:

$$|Z| = \sqrt{R^2 + \omega^2 L^2}, \tag{5.3}$$

$$\varphi = \tan^{-1} \frac{\omega L}{R}. \tag{5.4}$$

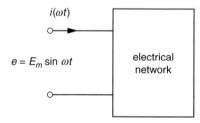

Fig. 5.4 Basic single-phase load.

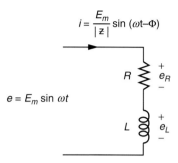

Fig. 5.5 Single-phase series R–L circuit.

For a resistance–inductance load, the time variations of current physically lag the variations of voltage and the circuit convention is that the time-phase angle ϕ is then negative in Eq. (5.2).

In terms of instantaneous current variation in Fig. 5.5, the circuit voltage may be written:

$$e(\omega t) = iR + L\frac{di}{dt}. \tag{5.5}$$

By Kirchhoff's loop law, the instantaneous total voltage is the sum of the component voltages e_R and e_L across the separate load components R and L, respectively:

$$e_R = iR = \frac{E_m R}{|Z|}, \tag{5.6}$$

$$e_R = L\frac{di}{dt} = \frac{E_m \omega L}{|Z|}\cos(\omega t - \varphi), \tag{5.7}$$

where

$$e = e_R + e_L. \tag{5.8}$$

The root mean square (rms) or effective value of a time-varying voltage $e(\omega t)$ that is repetitive in 2π radians is defined as:

$$E = \sqrt{\frac{1}{2\pi}\int_0^{2\pi} e^2(\omega t)\,d\omega t}. \tag{5.9}$$

For the particular case of sinusoidal voltage, if Eq. (5.1) is substituted into Eq. (5.9), it is found that:

$$E = \frac{E_m}{\sqrt{2}}. \tag{5.10}$$

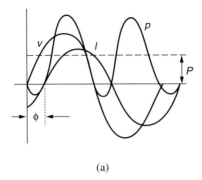

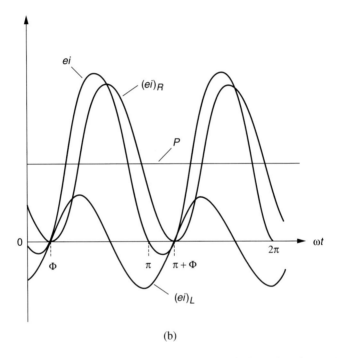

Fig. 5.6 Time variation in a series R–L circuit with sinusoidal supply voltage. (a) voltage, current and instantaneous power (b) analytical components of instantaneous power.[4]

The rms or effective value I of the current, in amperes (amps), of an instantaneous function $i(\omega t)$ that is mathematically continuous and periodic in 2π is defined by the basic expression.

$$I = \sqrt{\frac{1}{2\pi} \int_0^{2\pi} i^2(\omega t) d\omega t}. \tag{5.11}$$

It is seen that Eq. (5.11) corresponds to the definition of rms voltage in Eq. (5.9). For the case of sinusoidal current, the rms value I is related to the maximum value I_m of the current sinusoid by:

$$I = \frac{I_m}{\sqrt{2}} = \frac{1}{\sqrt{2}} \frac{E_m}{|Z|}. \tag{5.12}$$

Effective voltage E and effective current I are the values that define the circuit power dissipation. They are the values that would be read on properly connected moving-iron (i.e. standard AC) instruments. In calculations on circuits with sinusoidal supply voltage and sinusoidal current of the same frequency, the voltage and current are often represented in phasor form and calculations made by reference to phasor diagrams.

5.3 Power Relationships in Sinusoidal Circuits[4]

5.3.1 *Instantaneous power*

The product of instantaneous terminal voltage $e(\omega t)$ and instantaneous current $i(\omega t)$ gives a quantity known as the instantaneous power, $p(\omega t)$:

$$p(\omega t) = e(\omega t) \times i(\omega t). \tag{5.13}$$

For the circuit of Fig. 5.1, the instantaneous power, which is the instantaneous rate of energy transfer, is obtained by substituting Eqs. (5.1) and (5.2) into Eq. (5.13).

$$p(\omega t) = e \times i = \frac{E^2}{|Z|} \sin(\omega t) \sin(\omega t - \varphi) \tag{5.14a}$$

$$= \frac{E^2}{|Z|} \cos\varphi(1 - \cos(2\omega t)) - \frac{E^2}{|Z|} \sin\varphi \sin(2\omega t) \tag{5.14b}$$

$$= \frac{E^2}{|Z|}[\cos\varphi - \cos(2\omega t - \varphi)]. \tag{5.14c}$$

Instantaneous time variations of $e(\omega t)$, $i(\omega t)$, and $p(\omega t)$ are shown in Fig. 5.6(a). It is seen that $p(\omega t)$ is a sinusoid of double frequency 2ω, compared with the identical frequencies ω of the voltage and current. The values of the instantaneous powers associated with the two impedance components in the circuit of Fig. 5.5 are found by combining Eq. (5.2) with Eqs. (5.6) and (5.7), respectively:

$$e_R i = (ei)_R = \frac{E^2}{|Z|} \cos\varphi [1 - \cos 2(\omega t - \varphi)], \quad (5.15)$$

$$e_L i = (ei)_L = \frac{E^2}{|Z|} \sin\varphi [\sin 2(\omega t - \varphi)]. \quad (5.16)$$

The component instantaneous powers $(ei)_R$ and $(ei)_L$ in Eqs. (5.15) and (5.16) sum to total instantaneous power ei at the circuit terminals.

$$ei = (ei)_R + (ei)_L. \quad (5.17)$$

Although the component rates of instantaneous energy transfer $(ei)_R$ and $(ei)_L$ sum to the total system transfer (5.17), it is important to note that the values of $(ei)_R$ and $(ei)_L$, in Eqs. (5.15) and (5.16), respectively, do not coincide with the component terms of ei. The component expressions for total instantaneous voltamperes in Eq. (5.14) give no indication of the actual distribution of instantaneous voltamperes between the circuit components. This fact is found to be true of most circuits.

The values of the terms, $(ei)_R$ and $(ei)_L$ in Eqs. (5.15) and (5.16) obtained from the actual instantaneous current and voltages, are real and have physical existence associated with their respective impedance components. The time-variant terms on the right-hand side of Eq. (5.14), however, are only mathematical parts of a whole. They have no physical reality and no physical significance in relation to the impedance components of the circuit.

Time variations of the physical components $(ei)_R$, $(ei)_L$, and (ei) are shown in Fig. 5.6(b), for the case of a lagging (i.e. inductive) phase-angle φ. All three components are seen to vary sinusoidally at a frequency 2ω, where ω is the supply angular frequency in radians per second.

5.3.2 Average power and apparent power

The time average value P of the time-varying instantaneous power is called the real power, active power, or average power. For a mathematically continuous function $ei(\omega t)$ that is periodic in 2π, the average value P is defined by:

$$P = \frac{1}{2\pi} \int_0^{2\pi} ei \, d(\omega t). \quad (5.18)$$

This is the value that would be read by a suitable wattmeter properly connected at the circuit terminals. It has the physical dimension of power and satisfies the Principle of Conservation of Energy. The total power value of any number of elements connected in the circuit sum to the terminal value of the power. The reader should note that the definition of average power (5.18) is independent of waveform and frequency. It applies not only to sinusoidal circuits but also to the many nonsinusoidal circuits that occur in electronics.

For the particular case of a series resistance–inductance circuit from a sinusoidal supply (Fig. 5.2), the average power P can be found by substituting $ei(\omega t)$ from Eq. (5.14) into Eq. (5.18) and integrating. It is found that the time average value of the power is given by:

$$P = \frac{E^2}{|Z|} \cos \varphi = EI \cos \varphi. \tag{5.19}$$

This result can also be deduced directly from the relationship (5.14)(b) and (5.14)(c). The time variant parts of these two relationships, being sinusoidal or cosinusoidal, have time average values of zero, leaving the value shown in Eq. (5.19), In Fig. 5.6, the constant (time invariant) value of average power P is shown as a horizontal line.

The product of rms voltage E and rms current I is called the apparent power S.

$$S = EI. \tag{5.20}$$

Apparent power S has the dimension of volts times amps, which is "voltamperes". This is not the same value as the real (average) power in watts, except for the case of purely resistive circuits.

The apparent power values of the components of a circuit do not, in general, follow any conservation principles. For example, the apparent powers of the resistance and inductance in Fig. 5.5 do not sum to the terminal value of the apparent power. This is because, unlike the real power P, the apparent power S does not have a physical nature but is only a mathematical quantity.

5.3.3 Power factor

The energy efficiency of a circuit seen from its terminals is often defined in terms of a property known as the "power factor", which is the ratio of the average power P entering a circuit to the apparent power $S = EI$ at the terminals. By universal definition, the power factor is "the factor by which the apparent power must be multiplied to obtain the average power".

$$\text{Power Factor} = \frac{\text{average power}}{\text{apparent power}}. \quad (5.21)$$

In terms of the symbols defined in this section, it may be written as:

$$\text{PF} = \frac{P}{S} = \frac{P}{EI}. \quad (5.22)$$

The power factor is a dimensionless factor that has a fractional value within the range $0 \leq \text{PF} \leq 1$. Its maximum realisable value of unity can only be attained in circuits with sinusoidal supply voltage containing load impedances that are resistive or are compensated to be effectively resistive. It should be noted that the definition of power factor in Eq. (5.22) is independent of current or voltage waveforms because the definitions of P (5.18), E (5.9), and I (5.11) are all independent of waveform. Equation (5.22) applies to all circuits supplied by a sinusoidal supply voltage including circuits with nonsinusoidal current that occurs with circuits containing power electronic switching devices.

In electrical supply systems the rms voltage is usually substantially constant. This means that the power factor is proportional to the ratio of average power to rms current I.

$$\text{PF} \propto \frac{P}{I}. \quad (5.23)$$

The most energy-efficient condition of operation is when any stipulated load power P is supplied using the minimum possible value of current I. This is the condition of unity power factor ($\text{PF} = 1$). With minimum load current, there is then minimum power loss in the conductors of the transmission system and therefore minimum cost to the generating system.

To obtain the highly desired condition of unity power factor in a circuit, it is necessary that three conditions simultaneously be fulfilled. These are that the instantaneous terminal voltage and instantaneous terminal current must simultaneously:

1. have the same frequency.
2. have the same waveform (whatever that waveform may be, not necessarily sinusoidal).
3. be in time-phase at every instant of the cycle.

Each of these conditions is a necessity. If any one of them fails, the condition of unity power factor cannot be achieved. Except for the restricted case of circuits with purely resistive load, the circuit power factor is lower than unity because the apparent power S is greater in magnitude than the average power P. In electrical power supply systems, the generated voltages are usually sinusoidal. For most loads, both domestic and industrial, sinusoidal currents are delivered to the loads. With sinusoidal operation, the voltages and currents have the same frequency and waveform, thereby satisfying the first two conditions above.

With resistive loads, the voltage and current are in time phase at every instant of the cycle, so satisfying condition 3 above and verifying the unity power factor state. With reactive loads, however, inductive or capacitive, there is a time-phase difference between the voltage and current so that condition 3 is not satisfied. Any circuit that contains inductance or/and capacitance in its impedance such as an electric motor, must necessarily operate at less than unity power factor.

5.3.4 Reactive power

In sinusoidal circuits describing electrical power systems used throughout this book, the analytical difference between the apparent power $S = EI$ and the average (real) power P can be expressed in terms of a quantity known as the reactive power Q.

$$Q = EI \sin \varphi = \frac{E^2}{|Z|} \sin \varphi. \qquad (5.24)$$

The expression $E^2/|Z| \sin \varphi$ in Eq. (5.24) is seen to be a term in Eq. (5.14) (b), being one of the forms of representation of the instantaneous power $p(\omega t)$. In Eq. (5.14) (b), the value Q represents the peak amplitude of a time oscillatory term of double-supply frequency. It can be seen that the reactive power expression (5.24) is the analytical dual of the average (real) power P in Eq. (5.19). Comparing the expressions (5.19), (5.20), and (5.24), it is seen that:

$$P^2 + Q^2 = E^2 I^2 (\cos^2 \varphi + \sin^2 \varphi) = E^2 I^2 = S^2. \qquad (5.25)$$

The magnitudes of the average power P, the reactive power Q, and the apparent power S, from Eq. (5.25) can be used to represent the three sides

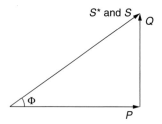

Fig. 5.7 Power triangle for a series R–L circuit of phase-angle Φ.

of a right-angled triangle in which S is the hypotenuse (Fig. 5.7). This figure is also appropriate to the components of power entering a series resistance–inductance circuit, in which the reactive power Q is represented as a positive quantity. The triangular geometrical representation of the components P, Q, and S is closely related to the form of the corresponding phasor representation of the voltages and currents of the circuit but is not itself a phasor diagram.

With sinusoidal circuits, for any topology or any number of circuit components, the reactive power is entirely associated with energy storage or retrieval. Any circuit can be reduced to a series resistance–inductance or resistance–capacitance equivalent in which the circuit reactive power is entirely associated with the circuit inductance or capacitance element. This is represented in Fig. 5.6 (b) where the instantaneous power $(ei)_L$ associated with the inductor L is a double-supply frequency sinusoid representing an oscillation of energy flow between the power supply and the magnetic field of the load inductor. The peak value of the double frequency oscillation of $(ei)_L$ is the reactive power Q. Equation (5.25) may also be expressed as:

$$\left[\begin{array}{c}\text{apparent}\\ \text{power}\end{array}\right]^2 = \left[\begin{array}{c}\text{average}\\ \text{power}\end{array}\right]^2 + \left[\begin{array}{c}\text{peak amplitude}\\ \text{of the instantaneous}\\ \text{reactive power}\end{array}\right]^2. \quad (5.26)$$

Electrical induction motors absorb not only real power P, to supply the mechanical power delivered via their shafts, but also lagging reactive power Q, to energise their magnetic fields. The effect of the reactive power consumed is to cause induction motors to operate at a lagging power factor that varies with motor speed and load.

An electrical supply utility usually has the task supplying its consumers with both real power (average power) P in watts and lagging reactive power Q in voltamperes. The implications of this involve not only the engineering considerations such as the electrical ratings of the supply network and its components (e.g. transformers, switchgear, etc.), but also the vital cost considerations such as the installation costs and the tariffs to be paid by the consumers for electricity supply.

5.4 Complex Power[4]

The real power P and reactive power Q components of the apparent power S can be interpreted in terms of conjugates in the complex plane. Real power components are aligned along the real (reference) axis while reactive power components are on the perpendicular axis or "j" axis. This permits the apparent power to be represented in Cartesian geometrical terms often called "phasor notation".

In order to facilitate calculations, in which the real power is always positive but the reactive power may be positive or negative, it is necessary to use the concept of "complex" power.

Consider the phasor representation E, I, of the supply voltage and current of Fig. 5.4. At an arbitrary instant of time, the voltage phasor, of magnitude E, is at an angle α with respect to the datum axis and the current phasor, of magnitude I_s, is at an angle β with respect to the datum axis (Fig. 5.8).

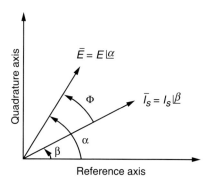

Fig. 5.8 Phasor diagram of voltage and current for single-phase circuit with inductive load.

The voltage and current phasors can be represented in Cartesian form:

$$\bar{E} = E_r + jE_q, \tag{5.27}$$

$$\bar{I}_s = I_r + jI_q. \tag{5.28}$$

The magnitude of the phasors is equal to the rms or effective values of the variables.

$$|\bar{E}| = E = \sqrt{E_r^2 + E_q^2}, \tag{5.29}$$

$$|\bar{I}_s| = I = \sqrt{I_r^2 + I_q^2}. \tag{5.30}$$

Consider the product $\overline{EI_s}$ of the two phasors E and I in Fig. 5.8.

$$\overline{EI_s} = (E_r + jE_q)(I_r + jI_q)$$
$$= (E_r I_r - E_q I_q) + j(E_r I_q + E_q I_r). \tag{5.31}$$

But the average power P and the reactive power Q supplied to the load of Fig. 5.4. can be represented, in terms of Fig. 5.8, as follows:

$$P_s = E I_s \cos\varphi$$
$$= E I_s \cos(\alpha - \beta)$$
$$= E_r I_r + E_q I_q. \tag{5.32}$$

$$Q_s = E I_s \sin\varphi$$
$$= E I_s \sin(\alpha - \beta)$$
$$= E_q I_r - E_r I_q. \tag{5.33}$$

Comparison of Eqs. (5.31)–(5.33) shows that the product $\overline{EI_s}$ of the voltage and current phasors, (5.31), does not give a meaningful result in phasor terms.

Now, consider the phasor conjugates $\overline{E^*}$ and $\overline{I^*}$ of the voltage and current, defined below:

$$\overline{E^*} = E_r - jE_q^*, \tag{5.34}$$

$$\overline{I^*} = I_r - jI_q^*. \tag{5.35}$$

It is found that the product $\overline{E^*I^*}$ of the conjugates also produces an anomalous result, similar to that of Eq. (5.31) in which once again, the component terms have no physical meaning. In order to obtain a product of voltage and current terms that produce a result consistent with the P and Q terms of Eqs. (5.32) and (5.33), it is necessary to define the quantity known as the complex power S^*. Two alternative, arbitrary definitions are used in the literature, as described below.

Let:

$$S^* = \overline{E^*I_s}$$
$$= (E_r - jE_q)(I_r + jI_q)$$
$$= (E_rI_r + E_qI_q) - j(E_qI_r - E_rI_q). \quad (5.36)$$

Alternatively, let:

$$S^* = \overline{E\,I_s^*}$$
$$= (E_r + jE_q)(I_r - jI_q)$$
$$= (E_rI_r + E_qI_q) + j(E_qI_r - E_rI_q). \quad (5.37)$$

Comparing Eqs. (5.36) and (5.37) with Eqs. (5.32) and (5.33) shows that:

$$\overline{E\,I_s^*} = P + jQ, \quad (5.38)$$
$$\overline{E^*I_s} = P - jQ. \quad (5.39)$$

In connection with the distribution of average power P and reactive power Q in linear and sinusoidal systems, the concept of complex power S^* is useful because it permits separate algebraic summation of both P and Q at any point in the system. This summation is possible because in linear and sinusoidal systems — and only in linear and sinusoidal systems — the reactive power is associated with energy storage. In linear sinusoidal systems, only the fundamental (i.e. supply frequency) component of current exists. The power factor $\cos\varphi$ is concerned entirely with energy storage.

In a diagrammatic representation of P, Q, and S^* as in Fig. 5.7, the angle between S^* and P is equal to the phase angle between the corresponding phasor components of voltage and current in Fig. 5.8. But it should be noted that the geometrical diagram of Fig. 5.7 is not a phasor diagram because the average power P and reactive power Q are quite different in nature.

Moreover, the time pulsations of P and Q are of double-supply frequency so that Figs. 5.7 and 5.8 could not legitimately be placed in the same diagram.

The concept of complex power S^* is entirely mathematical. It has the dimension of voltamperes but no physical existence and it cannot therefore be measured. Its magnitude can be determined by the separate measurement of P and Q and use of the relation:

$$|S^*| = \sqrt{P^2 + Q^2}. \qquad (5.40)$$

The case shown in Fig. 5.8, in which the voltage phasor leads the current phasor in time phase, represents current flow into an inductive load and is directly consistent with Eq. (5.38). Accordingly, the definition of Eq. (5.37) is the definition of complex electric power recommended and used by electric power utilities in the UK. By this definition, the reactive power supplied to an inductive load is considered positive and the reactive power supplied to a capacitive load is negative. This convention $S^* = \overline{EI_s^*}$ means that positive reactive voltamperes (i.e. lagging VArs) are generated by overexcited synchronous generators, capacitors, cables, and lightly loaded transmission lines and supplied to underexcited synchronous motors, induction motors, transformers, and inductance loads such as heavily loaded transmission lines. This convention is clearly oriented to the viewpoint of the electricity supply industry, which is in the business of generating and selling real (average) power in watts and (mainly) lagging power.

The convention for complex power $S^* = \overline{E^*I_s}$ in Eq. (5.36) is the one that is recommended by the International Electrotechnical Convention (EIC) and this leads to the quantity $-(E_q I_r - E_r I_q)$ being defined as positive reactive voltamperes. In EIC terminology, the quantity $S^* = EI^*$, in Fig. 5.7 for example, is known as the conjugate complex power.

References on power and reactive voltampere distribution in sinusoidal systems are divided with regard to the merits of the EIC convention. Most references originating in the USA support the convention while most references originating in the UK prefer the alternative. The reader is cautioned that when consulting references concerning reactive power, it is always wise to first ascertain the author's sign convention.

It is important to note that the concept of complex power is meaningless in nonsinusoidal systems. If the supply voltage is nonsinusoidal, or the load is nonlinear, or both, then phasor representation of voltage and current is invalid and resolution of the total power into Cartesian components is mathematically invalid as well as being physically meaningless.

Apparent voltamperes, being the product of rms supply voltage and rms supply current, are the figure of merit that define the maximum energy transfer capability of the system. Its definition is independent of waveform and it can therefore be used in discussions concerning power factor and power factor correction in nonsinusoidal systems as well as in linear and sinusoidal systems.

The apparent power S and the complex power S^* are equal in magnitude irrespective of sign conventions:

$$|S| = |E\,I| = |S^*|. \tag{5.41}$$

While the apparent power S is a scalar quantity, possessing only magnitude, the complex power S^* can be interpreted, in sinusoidal circuits and systems only, as having Cartesian coordinates equal to the average power P and the reactive power Q.

5.5 Real Power Flow and Reactive Power Flow in Electrical Power Systems

5.5.1 *General summary*

The various load components in an electrical transmission and distribution system absorb both real average power P and reactive power Q. Components of the system itself such as transformers, switchgear, power factor correction reactors, and protection devices also absorb both real and reactive powers.

All of the powers, both real and reactive, are supplied by generators and static reactor equipment within the system. The entire interconnection of source units and load units is referred to as an electrical grid or as a network or system. Wind-energy generators are usually linked into a grid and operate in parallel with other generators, rather than operating independently. When a large number of generators are linked together the system has a large electrical inertia — the system voltage and frequency are kept effectively constant at levels prescribed by law. The system is then referred to as an "infinite bus" (short for "infinite busbars"). Any individual component, whether a source unit or a load unit, must operate at the bus voltage and frequency. Reactive power is associated with energy flow (but not energy dissipation) in fields of force. When the field of force is a magnetic field, as in electrical rotating machines and transformers, the circuit representation is in terms of inductance. The inductance parameter L that represents a magnetic field system is proportional to the magnetic energy stored in the field. When the field of force is an electric field, as in power factor correction equipment, the circuit representation is in the form of capacitance. The capacitance parameter C that represents an electric field system is proportional to the electric energy stored in the field.

In order to accommodate both the magnetic type and the electric type of reactance and reactive power in power systems calculations, it is necessary to adopt a sign convention. The convention invariably used is that magnetic (inductive) reactive power absorbed is positive while electric (capacitive) reactive power absorbed is negative. The four quadrants of operation for load power absorbtion are shown in Fig. 5.9. When real power P and inductive reactive power Q are being absorbed, as in a transformer, the circuit representation is a series R–L circuit, as in quadrant I. When power P and inductive reactive power Q are being absorbed, as in power factor correction equipment, the circuit representation is an series R–C circuit, shown in quadrant IV.

Operations represented in quadrant II and quadrant III in Fig. 5.6 take place in the negative real power regions. If positive real power is power

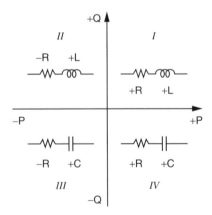

Fig. 5.9 Circuit representation of real power P and reactive power Q absorbtion in power systems.

absorbed by a load, then negative real power implies power delivered to the system (not absorbed by it). It follows that the real power P in quadrants II and III is representative of generation, not of load power absorption.

5.5.2 Summary from the perspective of the consumer[5]

1. The consumer absorbs both P and Q but usually is metered (charged) only for P.
2. The utility has to supply not only the customers demanded power P but also the power losses in the transmission system conductor resistances.
3. The load voltage level is particularly sensitive to increments in Q rather than increments in P.
4. In order to minimise the transmitted power P (and the fuel used to generate it) and to maintain the stability of the load voltage, the utility encourages customers to operate at the maximum possible power factor. This implies that consumers will draw their required P at the minimum possible current level. This further implies that the consumer Q demand should be the minimum possible, ideally zero.
5. Special tariffs are applied by the utilities that penalise consumers for large Q demand (i.e. for operating at low power factor).
6. Many consumers can minimise the cost of their electricity supply through the installation of power factor capacitors.

References

1. Weedy, B. M., *Electric Power Systems*, Third Edition (revised), John Wiley & Sons, New York, the USA, 1987.
2. Toro, V. D., *Electric Machines and Power Systems*, Prentice Hall Inc., Englewood Cliffs, N.J., the USA, 1985.
3. Wildi, T., *Electrical Machines, Drives and Power Systems*, Prentice Hall Inc., Upper Saddle River, N.J., the USA, 2000.
4. Shepherd, W., P. Zand, *Energy Flow and Power Factor in Nonsinusoidal Circuits*, Cambridge University Press, London, England, 1979.
5. Freris, L. L. (Ed.), *Wind Energy Conversion Systems*, Prentice Hall International (UK) Ltd., Englewood Cliffs, N.J., the USA, 1990.

CHAPTER 6

Electrical Generator Machines in Wind-Energy Systems

In most electrical power systems, the bulk of the power is supplied by rotating machine generators. These can be of many types, but those used for wind-power generation are mainly the following:

- direct current (DC) machines,
- alternating current (AC) synchronous machines, and
- induction machines.

6.1 DC Generators

Small stand-alone wind generator systems were at one time fitted with shunt-wound DC generators. This form of self-excited generator is easy for speed control and was used extensively until the early 1980s, especially when the output power could be used in DC form. The need for a machine commutator and brush gear resulted in low reliability and high maintenance costs.

Modern brushless DC machines use permanent magnet excitation to eliminate the need for a machine commutator. This configuration essentially forms an inside-out conventional DC machine with its mechanical commutator replaced by an electronic commutation circuit. This brushless machine technology greatly improves machine reliability but is restricted to machine ratings of less than a few hundred watts at most. As such, it is not suitable for the multi-megawatt machines installed in modern wind-energy generation schemes.

6.2 AC Generators

Most electricity supply systems are of three-phase AC form with fixed frequency and voltage, and use sinusoidal voltages and currents. Electrical power for modern transmission and distribution systems is invariably generated by three-phase, synchronous generators of many megawatts ratings. The largest of such machine sets in the UK are rated at 600 MW and are housed, with their fossil or nuclear fuelled turbines, in large utility power plants or power stations.

Wind-energy generators are several orders smaller in size, but are increasing in rating all the time. In the mid-1990s, the largest wind turbine was the Nordex N52 machine rated at 800 kW, although the average size was about 250 kW. Ten years later, in 2004, the largest machines were in the range of 3–5 MW. Over the same 10 year period, wind turbine rotor diameters increased from 7.44 to 61–90 m.

AC generators in wind power systems fall into two main categories.

1. Synchronous generators, requiring DC winding excitation or permanent magnet excitation, that deliver a voltage and frequency which are both proportional to the rotation speed.
2. Induction generators, requiring AC excitation and the frequencies and magnitudes of the generated voltages depend both on the frequency of the excitation current and on the speed and direction of rotation.

Synchronous generators and induction generators are both very important in wind-energy systems. Each of these is covered, in some detail, in following sections.

6.3 Synchronous Machine Generators[2,3]

A three-phase synchronous machine is depicted in Fig. 6.1. Its stator, also called the armature, consists of iron cores with three groups of coils wound

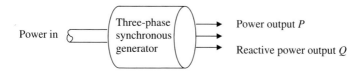

Fig. 6.1 General form of a three-phase synchronous generator.

around them. These identical coils, also named three "phase" windings, have 120°-phase shift between each of them and are connected directly to a three-phase power supply system or to a load when standing alone. In this text, only the former is considered. Coils wound on the rotor core of a three-phase synchronous generator are energised by a DC voltage source. This sets up a field of magnetic flux which cuts the armature conductors as it rotates. The individual coils are connected in a manner that provides alternate north and south poles. Relative motion between the rotating field flux and the armature conductors causes electromotive forces (emf), i.e. voltage, in the stator coils. Detailed descriptions of synchronous machine construction and performance are provided in many existing books on electrical machines and are not included in the present text.

With a wind turbine as the prime mover, power flows into the machine characterised by the torque T on the rotor shaft and the speed of rotation. In SI units:

$$P_{in} = T\omega_s. \tag{6.1}$$

The power, P_{in}, is in watts when the shaft torque T is in newton metres and the shaft rotational speed ω_S is in radians/second.

All of the real power, P_G, generated by the machine comes in through the shaft from the turbine, which also provides all the real power losses in the machine stator and rotor windings.

A synchronous generator connected into a power supply system can be represented by the single-phase equivalent circuit shown in Fig. 6.2. The generator frequency is determined by the rotor's rotational speed and the number of poles as:

$$f = \frac{P_k \cdot N_s}{120}, \tag{6.2}$$

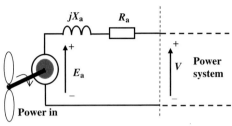

E_a: per-phase generated voltage, V
R_a: per-phase armature resistance, Ω
X_a: per-phase synchronous reactance, Ω
I_a: per-phase armature current, A
V: per-phase terminal voltage, V

Fig. 6.2 Single-phase equivalent circuit of a three-phase synchronous generator.

where f is the frequency in hertz (Hz), P_k is the number of rotor magnetic poles, and N_s represents the shaft synchronous speed in rev/min (rpm). P_k is a design constant so that the frequency of the generated voltage is directly proportional to the machine rotating speed.

The desired frequency of generation is usually 50 or 60 Hz. In order to generate at a frequency of 60 Hz, for example, the necessary speed of shaft rotation for a range of pole numbers is shown in Table 6.1 calculated from Eq. (6.2).

For a generator connected to a large power system, depicted in Fig. 6.2, the generator speed of rotation, N_S, must be governed to remain constant at the value consistent with the system frequency.

Table 6.1 Synchronous speed of a synchronous machine or induction machine.

At 60 Hz								
No. of magnetic poles, P_k	2	4	6	8	......	72	144	288
Synchronous speed (rpm)	3600	1800	1200	900		100	50	25
At 50 Hz								
No. of magnetic poles, P_k	2	4	6	8	......	72	144	288
Synchronous speed (rpm)	3000	1500	1000	750		83.3	41.67	20.83

With fixed rotational speed, the stator induced voltage E_a, in Fig. 6.2, is derived from the DC excitation winding, not shown, and is adjustable in value by field current control. Each phase winding of the generator armature may be characterised by the winding resistance R_a and an impedance element known as the "synchronous reactance", X_a. This is not a real physical element but an analytical component arising from sinusoidal flux distribution in the armature winding and magnetic cores of the machine. The currents in the stator phase windings will set up magneto motive forces (mmfs) and therefore flux linkages; hence, the windings must possess self- and mutual inductances across which voltage will be dropped. In the presence of load impedance — in this case the coupled power system — currents I_a flow in the armature windings. The per-phase voltage V at the generator terminals is fixed at the system value and is related to E_a by the phasor equation:

$$V = E_a - I_a(R_a + jX_a). \tag{6.3}$$

The use of 'j' notation is valid only for quantities varying sinusoidally in time.

Adjustment and control of voltage E_a, in Fig. 6.2, controls the reactive power generated. The action of changing the field current and its effect on generator operation is best illustrated in terms of phasor diagrams shown in Fig. 6.3. This demonstrates the condition where an overexcited cylindrical rotor generator delivers current to the system that lags the terminal voltage V in time phase by the angle Φ_a. Usually the voltage drop $I_a X_a$ is at least 10 times the voltage drop $I_a R_a$, so that often the latter is neglected altogether. In the phasor diagram, the voltage component $I_a R_a$ is in time phase with current I_a, while the component $jI_a X_a$ leads I_a in time phase by 90°. The

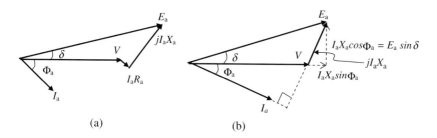

Fig. 6.3 Phasor diagram of an overexcited synchronous generator: (a) delivering a lagging current and (b) delivering a lagging current – neglecting armature resistance.

angle δ between the voltage E_a and V is usually referred to as "power angle", "load angle", or "torque angle".

If the armature resistance is neglected, the phasor diagram of Fig. 6.3(a) can be reinterpreted in more detail in Fig. 6.3(b). The real power P delivered by the three-phase generator is given by:

$$P = 3VI_a \cos \Phi_a. \qquad (6.4)$$

The reactive power delivered by the generator is given by:

$$Q = 3VI_a \sin \Phi_a. \qquad (6.5)$$

In Fig. 6.3(b), it can be seen that:

$$E_a \sin \delta = I_a X_a \cos \Phi_a. \qquad (6.6)$$

The term I_a, $\cos \Phi_a$ can be eliminated between Eqs. (6.4) and (6.6) to give the expression:

$$P = \frac{3VE_a}{X_a} \sin \delta. \qquad (6.7)$$

Similarly, the reactive power Q out of the generator has the magnitude:

$$Q = \frac{3V}{X_a}[E_a \cos \delta - V]. \qquad (6.8)$$

When the generator is connected to an infinite bus bar, the terminal voltage V is constant. Although the value of the synchronous reactance X_a depends on the state of the generator magnetic saturation, it is almost constant within the normal range of generator operation. From Eq. (6.4):

$$P \propto I_a \cos \Phi_a \qquad (6.9)$$

and from Eq. (6.7):

$$P \propto E_a \sin \delta. \qquad (6.10)$$

An increase of the output power P delivered by the generator is realisable by increasing variables E_a, I_a, $\cos\Phi_a$, and $\sin \delta$. But these four variables are interactive — an increase of one will cause a decrease of another. For example, Fig. 6.4 shows the effect of increasing E_a when the torque angle δ is constant. It is seen that the phase $I'_a X_a$ increases to $I''_a X_a$ representing an increase of the armature current delivered from I'_a to I''_a. But this increase is accompanied by an increase of the phase angle from Φ'_a to Φ''_a and a consequent decrease of $\cos \Phi'_a$ to $\cos \Phi''_a$. Concurrent increase of the

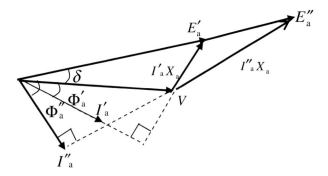

Fig. 6.4 Overexcited synchronous generator showing the effect of increase of the excitation voltage E_a.

delivered current I_a and the phase-angle Φ_a is seen from Eq. (6.5) to result in the increased delivery of lagging reactive power Q. Any change of the state of operation of a generator "on line" must be carefully considered to include the effects on all of the machine electrical variables. In particular, it is necessary to know the effect of any changes on the real power P and the reactive power Q delivered by the machine.

When the magnitude of the induced voltage E_a is smaller than that of the terminal voltage $|V|$, the effect is for the generator to deliver a leading current to the bus, Fig. 6.5. The leading voltamperes delivered to the system can only be absorbed by capacitance within the system, such as that provided by power factor correction capacitors. In effect, the delivery of a leading current from the generator to the power system is equivalent, in terms of reactive power flow, to imposing a lagging (inductive) load. An underexcited synchronous generator is equivalent to a sink of lagging

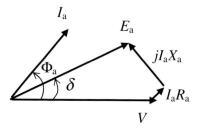

Fig. 6.5 Underexcited synchronous generator supplying leading current.

voltamperes, although it is actually delivering a leading current and is a source of leading voltamperes. Both the real power P and reactive power Q are invariant with machine speed. Load-flow analysis and transient effects are often considered in terms of variations of the load angle δ.

A three-phase synchronous machine connected to a large system can also operate as a motor, with constant terminal voltage. Because the operating frequency is also constant, at the bus value, the speed of rotation is fixed as described in expression (6.2). The number of magnetic poles P_k must be an even number and the speed of rotation is therefore fixed at some level described in Table 6.1. In order to realise a typical large size turbine with speed of (say) 25 rpm at 60 Hz, it would be necessary to design the machine for 288 poles which is entirely unrealistic.

Figure 6.6 shows that a synchronous motor is an energy converter with two separate electrical inputs and one mechanical output. Real power P is taken from the supply to provide the output power and the losses of the motor. The function of the field winding is to adjust the phase-angle of the current fed into the motor armature. With P constant, it is possible to vary the armature current I_a from a leading phase angle (with $|E_a| > |V|$), through unity power factor operation, to a lagging phase angle (when $|E_a| < |V|$). A synchronous motor is very versatile and can perform two functions simultaneously:

1. It can deliver mechanical and rotational powers to a load at its constant (synchronous) speed.
2. It can act as a sink of adjustable voltamperes that can be made to vary from lagging (inductive), through in-phase, to leading (capacitive) power factor.

The property of a synchronous motor in being able to absorb leading voltamperes is particular useful because it then acts like an adjustable capacitor. This property can be used for power factor correction in power systems and is especially useful to compensate the lagging power factors

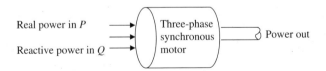

Fig. 6.6 General form of a three-phase synchronous motor.

caused by parallel connected induction motors. An overexcited synchronous motor is sometimes described as a "synchronous capacitor".

6.4 Three-Phase Induction Machine[2-4]

A three-phase induction machine, whether acting as a motor or a generator, has a double cylindrical structure (Fig. 6.7). A cylindrical rotor, containing distributed three-phase winding or a cage structure winding, is free to rotate at speed N, inside the cylindrical cavity of a stator structure built onto the stationary frame of the machine. As in a synchronous machine the stator has a distributed and three-phase winding. The two three-phase windings, on rotor and stator, are linked magnetically but not electrically. When a set of balanced and three-phase sinusoidal voltages is applied to the three-phase stator windings, the effect is to produce a "rotating magnetic field", illustrated in Fig. 6.7. The field does not physically rotate but the magnetic effect is precisely the same as if there was a pair of magnetic poles rotating around the airgap at a synchronous speed N_s, depending on the frequency and the number of poles, defined by Eq. (6.2).

The three-phase induction machine acting as a motor is the workhorse of modern industry. Wherever rotation power is needed for machines, tools, appliances, systems, etc., this power is invariable provided by three-phase induction motors. It is likely that in the USA, for example, there are more three-phase induction motors than there are people. The ratio of three-phase

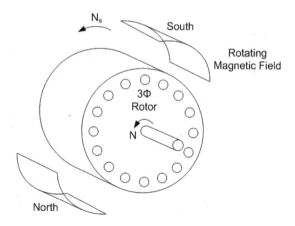

Fig. 6.7 Induction motor rotor acted on by a "rotating magnetic field".

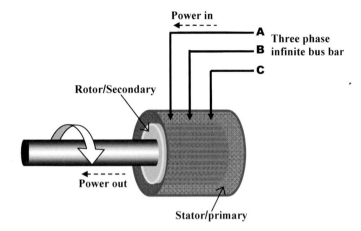

Fig. 6.8   General form of the three-phase induction motor.

induction motors to people is a rough measure of the level of industrialisation of a country.

Use of the three-phase induction machine as a generator is very much less common than its use as a three-phase motor. The three-phase induction generator is used in certain special industrial drives but is now mainly to be found in the greatly expanding wind-power industry.

An explanation of the action of the induction machine best begins by considering the operation of the three-phase induction motor. A consideration of generator is made later.

6.4.1 *Three-phase induction motor*

The three-phase winding on the machine stator is usually connected to a three-phase supply of constant voltage and frequency. This winding becomes the primary winding of the machine and also the excitation winding. Motor torque and output power is developed in the secondary windings on the machine rotor. These windings are closed on themselves directly, Fig. 6.8, or in the wound rotor case through equivalent resistors. External circuits can be connected into the motor secondary windings via slip-rings and brush gear on the rotor.

An induction motor operates by the principle of electromagnetic induction. The rotor conductors rotate in the direction of the stator rotating field at a velocity N_r that is lower than the synchronous rotating velocity N_s

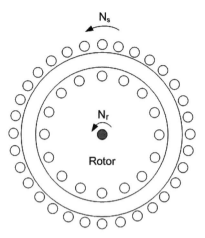

Fig. 6.9 Rotational velocity in the three-phase induction motor: N_s: stator flux rotating speed; N_r: rotor mechanical speed; SN_s: speed difference between stator flux and rotor; and $N_r + SN_s$: rotor flux with respect to stator.

(Fig. 6.9). The differential speed between the "rotating" stator flux N_s and the actual rotor speed N_r is called the slip speed.

$$\text{Slip speed} = N_s - N_r = SN_s. \quad (6.11)$$

The ratio of the slip speed to the synchronous speed is the most important variable in the induction motor operation and is referred to as the "per-unit slip".

$$S = \frac{N_s - N_r}{N_s} = 1 - \frac{N_r}{N_s}. \quad (6.12)$$

An alternative form of this relationship is the expression:

$$N_r = (1 - S)N_s. \quad (6.13)$$

If the rotor operates at synchronous speed, then $N_r = N_s$ and $S = 0$. At standstill $N_r = 0$ and $S = 1$. At reverse synchronous speed, which can happen during motor braking operation, $N_r = -N_s$ and $S = 2$.

Many induction motor characteristics, such as current variation, may be shown graphically as functions of motor speed. These can also be expressed as functions of slip, as shown in Fig. 6.10.

Because the rotor conductors cut the rotating flux field at the slip speed, the emfs induced in them are of slip frequency and the rotor currents that flow are also of slip frequency. If the supply (stator) frequency is f_s, then the corresponding rotor voltage/current frequency f_r, is given by:

$$f_r = Sf_s. \quad (6.14)$$

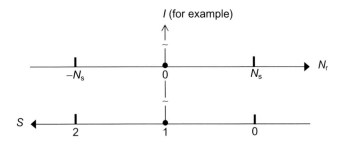

Fig. 6.10 Speed and slip expressed as alternate independent variables.

In Europe, the domestic frequency is usually $f_s = 50\,\text{Hz}$. In North America, the domestic electrical system operates at 60 Hz. The per-unit slip for full-load operation of a motor is usually about 5% or $S = 0.05$. This implies that the frequency of the rotor electrical variables at full load is 3 Hz, at $f_s = 60\,\text{Hz}$ and 2.5 Hz at $f_s = 50\,\text{Hz}$. The frequency of the rotor currents therefore varies from line frequency (50 or 60 Hz) at standstill to a low frequency of a few Hz at high speed. At synchronous speed, any rotor current is of zero frequency (i.e. DC).

Now, the magnetic fields of the primary (stator) and secondary (rotor) windings must be in synchronism with each other for torque to be generated. Since the stator flux rotates at synchronous speed N_s while the rotor rotates at speed N_r, this requires that the rotor field must rotate, with respect to the rotor, at speed $N_s - N_r$ or (SN_s) in the direction of rotation.

$$\text{Speed of rotation of the rotor flux with the stator} = N_r + (N_s - N_r)$$
$$= N_r + SN_s = N_s. \tag{6.15}$$

The flux components on the two windings therefore rotate in synchronism at N_s independently of the variable speed N_r of the motor shaft.

Because the motor works on the principle of electromagnetic induction, the magnitudes of the emfs induced in the secondary windings depend on the time rate of flux cutting (or linking). For a unity transformation ratio across the airgap (Fig. 6.11), the magnitude of the secondary emfs E_2, in terms of the primary emfs E_1, is:

$$E_2 = SE_1. \tag{6.16}$$

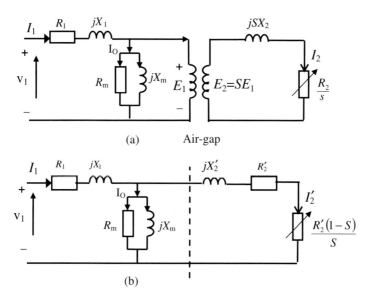

Fig. 6.11 Per-phase equivalent circuit of the three-phase Induction motor: (a) incorporating a transformation coupling and (b) referred to primary windings V_1: applied voltage, per phase; I_1: applied current, per phase; I_o: no-load current, per phase; I_2: secondary current, per phase; E_1: internal primary emf, per phase; E_2: internal secondary emf, per phase; R_1: internal primary winding resistance; X_1: internal primary leakage reactance; R_m: internal primary no-load resistance; X_m: internal primary magnetising reactance; R'_2: internal secondary winding resistance referred to primary; X'_2: internal secondary leakage reactance referred to primary; S: per-unit slip.

It is seen from Eqs. (6.14) and (6.16) that both the magnitude and frequency of the secondary induced emfs (and current) are proportional to the per-unit slip S.

The equivalent circuit approach in induction motor operation is complicated by the fact that the stator windings operate at synchronous (or line) frequency while the rotor operates at slip frequency. An accurate equivalent circuit has to contain a frequency transformation between the windings as well as a magnitude turns ratio representing transformer action. Both of these features are subsumed by using an equivalent circuit referred to primary turns (Fig. 6.11(b)). An appropriate change of secondary side parameters enables the transformation coupling between the primary and secondary windings in Fig. 6.11(a) to be eliminated, with regard to both the magnitude and frequency transformations.

In Fig. 6.11, the electrical power P_{ag} crossing the airgap is seen to be:

$$P_{ag} = 3|I_2|^2 \frac{R_2}{S} = 3|I_2|^2 \left(R_2 + \frac{R_2(1-S)}{S} \right). \quad (6.17)$$

Note the constant 3 represents three phases.

The power slip component in Eq. (6.17) is the electrical equivalent of the power converted into mechanical form plus the power to overcome rotational friction losses. This power can be expressed in terms of torque and speed. By comparison with the form of Eq. (6.1) the induction motor shaft torque "T" can be written as:

$$T = \frac{P_{ag}}{S\omega_s}. \quad (6.18)$$

Combining Eq. (6.18) with Eq. (6.13) gives:

$$P_{ag} = T\omega_r = T\omega_s(1-S). \quad (6.19)$$

The airgap power P_{ag} for an induction motor is seen to be the fraction $(1-S)$ of the synchronous input power $T_e N_s$ defined in Eq. (6.1). Substituting Eqs. (6.13) and (6.12) into Eq. (6.18) gives:

$$T = \frac{3|I_2|^2}{\omega_s} \frac{R_2}{S}. \quad (6.20)$$

Now the motor secondary current and its primary referred value I_r are not normally accessible for measurement. It is desirable to be able to express the torque in terms of the primary applied voltage, which is readily accessible. This can be achieved by using a slightly modified equivalent circuit (Fig. 6.12), which introduces a slight (few percent) error, compared with

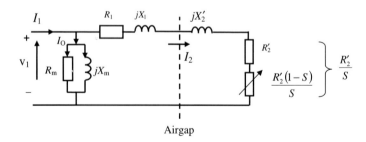

Fig. 6.12 Modified per-phase equivalent circuit of the three-phase induction motor.

Fig. 6.11. For the circuit of Fig. 6.12, it is seen that:

$$|I_2'| = \frac{V_1}{\sqrt{\left(R_1 + \frac{R_2'}{S}\right)^2 + (X_1 + X_2')^2}}. \quad (6.21)$$

Combining Eqs. (6.18) and (6.19) gives an expression for the motor torque in terms of measurable quantities.

$$T = \frac{3\frac{|V_1|^2}{\omega_s}\frac{R_2'}{S}}{\left(R_1 + \frac{R_2'}{S}\right)^2 + (X_1 + X_2')^2}. \quad (6.22)$$

Rewriting Eq. (6.22) in terms of the actual motor speed ω_r rather than the slip S does not render any analytical advantage and makes the expression more cumbersome.

A diagram of torque variation, for motoring operation between synchronous speed N_s and reverse synchronous speed, $-N_s$ is given is Fig. 6.13. Rated operation (i.e., rated torque at rated speed) usually occurs at about 95% forward speed ($S = 0.05$). In the region of quadrant IV, the speed is negative so that the motor-developed torque is acting against the rotation of the load, which represents braking operation.

6.4.2 Three-phase induction generator

An induction machine, described in detail in Sec. 6.4.1, for motoring operation, can equally well be used as a generator. The induction generator,

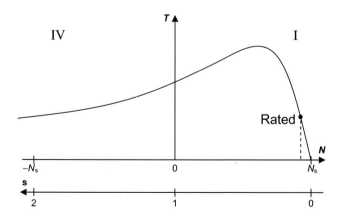

Fig. 6.13 Torque-speed characteristic for a three-phase induction motor.

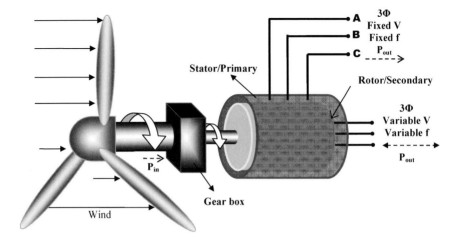

Fig. 6.14 General form of the three-phase induction generator.

used extensively in wind-energy generator schemes, is shown in general form in Fig. 6.14. Power passes into the machine through its shaft from the wind turbine. The machine stator windings are usually connected (directly or indirectly) to the three-phase power network and act as excitation windings. For squirrel-cage rotor machines, the stator connection also provides the route by which generated power is transferred into the power system and is called the singly fed induction generator (SFIG). In this form of connection, only one value of rotational speed can be adjusted so that its application is limited to constant speed wind turbines. Further discussion is described in Sec. 6.4.3. A more commonly used induction machine formation for wind-energy applications is the doubly fed induction generator (DFIG), which requires a wound-rotor machine and will be discussed in Sec. 6.4.4.

In order for electrical generation to occur with an induction generator, the rotational speed of the rotor N_r must exceed the rotational speed of the stator rotating field N_s. The slip speed SN_s in Eq. (6.11) is then negative and therefore the slip S is negative. Now, the rotating field of the stator excitation winding rotates forward at the synchronous speed N_s and the rotating field of rotor must be in synchronous with it. If the rotor speed N_r exceeds N_s, the rotor flux must rotate backwards at speed $N_s - N_r$ with respect to the rotor. The secondary emfs E_2, in terms of the primary emfs E_1, become negative; hence, the rotor current I_2 flows in the reverse direction to that in motoring mode.

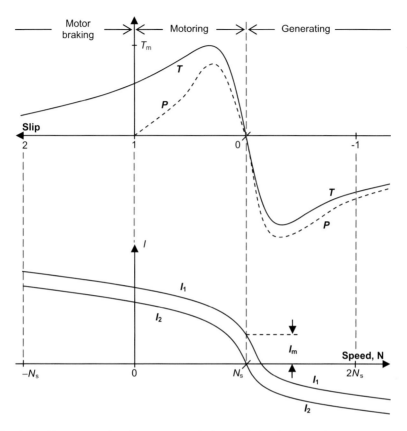

Fig. 6.15 Torque-speed and current-speed characteristics for motoring and generating operation of a three-phase induction machine.

The electrical power P_{ag} crossing the airgap, expressed in Eq. (6.17), is now negative from the rotor to the stator and is the electrical equivalent of the power converted from turbine mechanical form minus the power to overcome rotational friction losses. An overall torque-speed characteristic for both generating and motoring operation is given in Fig. 6.15. This subsumes the characteristic for motoring operation only that is shown in Fig. 6.13. It is shown for induction generator operation, that $N_r > N_s$ and $S < 0$.

The necessary corresponding values of synchronous speed are indicated in Table 6.1. At 50 Hz, a two-pole induction generator must operate at a speed exceeding 3000 rpm in the direction of rotating field. The developed torque is proportional to the slip speed $N_s - N_r$, which is negative in the working range for low-slip application. But the rated speed of the wind turbine is in the range a few tens of rpm for MW size generator and a few hundred rpm for kW size generators. In most practical cases, it is necessary to gear up the turbine shaft speed to the desired value.

For three-phase induction motoring operation, the machine always acts as a partially inductive load on the system. For all load conditions, it draws in real power to transmit to its load, via the rotor shaft, and also lagging reactive power. In equivalent circuit terms, Fig. 6.12 can be effectively reduced to a series resistance–inductance circuit with fixed inductance but variable resistance, depending upon the value of the slip. The necessary lagging vars are drawn from its electrical supply network.

When an induction machine operates as a three-phase generator, it still requires an input of lagging vars into its primary winding from the parallel connected generators of the electrical network. The secondary winding delivers real power, drawn from its mechanical input power, plus leading vars because the current delivered to the system is phase advanced with its phase voltage, as shown in Fig. 6.16. From the viewpoint of a power system load, an induction generator can be thought of as a sink of lagging vars operating at a lagging phase angle of 180-Φ as depicted in Fig. 6.16.

Since an induction generator delivers capacitive (leading) vars its load has to be capable of receiving leading vars. An individual induction generator, acting independently of any parallel connected sources can deliver its var load to a set of capacitors connected across its terminals (Fig. 6.17).[5,6] Although an external supply is not needed, the operating frequency and the generated voltage V_S are affected by the speed, the load, and the capacitor rating. The terminal capacitor must also supply any lagging

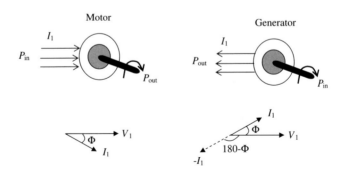

Fig. 6.16 Phase relationships of the induction motor and induction generator.

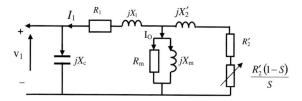

Fig. 6.17 Single-phase equivalent circuit of three-phase induction generator with compensating capacitors.

reactive power required by the external load. Provided that the induction machine rotor has an adequate remnant field the machine will self-excite. But for wind-generation applications, the condition of self-excitation is not usually desirable.

For a wind generator embedded into a power system in parallel with other synchronous and induction generators, its voltage and frequency are determined by the system bus bars and its lagging reactive power is supplied from the system. This is another way of saying that the leading vars delivered by the induction generators can be absorbed into other components of the network. In cases when an external load network has only limited ability to absorb the leading vars from the induction generators, local capacitors may be used as a supplement.

The usual power system practice is to regard induction generators as absorbers of lagging reactive voltamperes. The use of parallel-connected capacitance can be thought as a form of power factor correction.

The variation of real power P with speed is shown in Fig. 6.15. For generator operation, the machine losses have to be provided by the generator itself, rather than from the system as with motoring operation. Of main interest from the viewpoint of wind-energy conversion is the narrow speed range representing $\pm 30\%$. In this range, the variation of torque with slip is linear for both motoring and generating operation. Some idea of the power scale is shown in Fig. 6.18, quoted from Ref. 7, which represents a 1-MW induction machine. The normal operating range O-A represents a slip of -0.8%, which is 12 rpm for a 4-pole, 50-Hz machine and 14.4 rpm for a 4-pole, 60 Hz machine. Corresponding variation of Q is given in Figs. 6.19 and 6.20.[7] At point A in Fig. 6.19, which represents $P = 1$ MW of generator

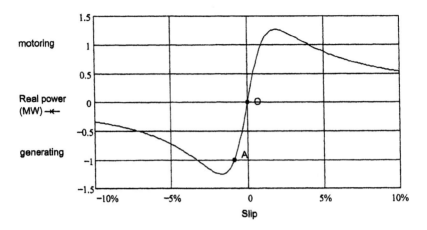

Fig. 6.18   Real power P versus slip for a 1 MW induction generator.[7]

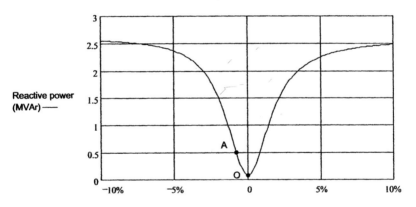

Fig. 6.19   Reactive power Q versus slip for a 1 MW induction generator.[7]

output, the machine draws a reactive power $Q = 600\,\text{kVAR}$ in the presence of its fixed local power factor correction capacitors.

6.4.3 Different generation systems

There are several different types of wind-power generation systems according to the induction generators used and the ways interfaced to the grid. The ones most widely used are discussed below.

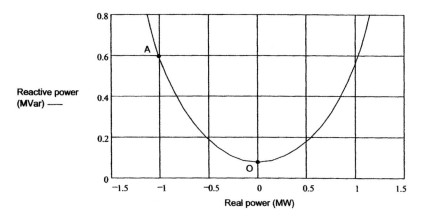

Fig. 6.20 Reactive power Q versus real power P for a 1 MW induction generator.[7]

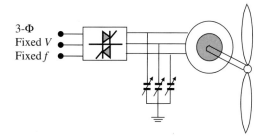

Fig. 6.21 Squirrel-cage induction generator for CSCF operation.

6.4.3.1 *Constant speed constant frequency (CSCF)*

The configuration of this system is a "Danish-concept" shown in Fig. 6.21. It consists of a squirrel-cage induction machine connected directly to the grid. The generator requires a voltage regulation circuit to prevent large inrush current and a capacitor bank for reactive power compensation. This is optimal at only one or two wind speeds. The power can be controlled only by blade pitch angle. Power factor cannot be controlled.

6.4.3.2 *Variable speed constant frequency (VSCF)*

1. A cage-rotor induction machine with its stator connected to the utility grid through a PWM DC-link converter.

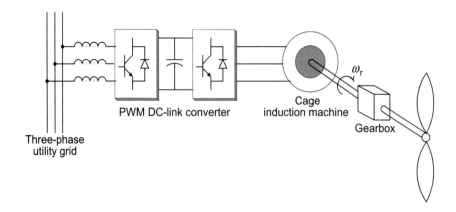

Fig. 6.22 VSG system using cage induction machine and PWM converters.

This system is commonly used for high-performance AC drives and is shown in Fig. 6.22.

The main advantage of this configuration lies in the simple and robust structure of the squirrel-cage induction machine. The system can operate with a wide range of wind speeds and permits start-up and regenerative braking modes. However, the converter rating needs to be approximately 125% of the machine rating, as the reactive power required by the machine must be supplied by the converter.

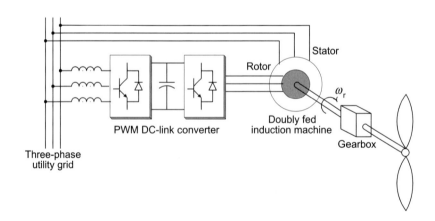

Fig. 6.23 VSG system using doubly fed induction machine and PWM converters.

2. A wound-rotor induction machine having its rotor connected to the grid through a PWM DC-link converter is called a doubly fed induction generator (DFIG).

This scheme may be regarded as a variation of the well-known Scherbius drive. A simplified block diagram of this system is shown in Fig. 6.23. In such a configuration, the converter rating required is only the slip-fraction of the machine power rating due to a restricted operating speed range. Moreover, the constant-frequency power is mainly generated from the machine stator giving sinusoidal currents supplied directly to the utility grid.

The power transfer relationship is as follows. The turbine mechanical driving torque, T_m, excluding machine frictional and windage losses, must balance the generator electromagnetic torque T_e at the steady state ($T_m \cong T_e$). The turbine mechanical power P_{mech} equals the sum of the power crossing the airgap P_{ag}, rotor power P_r plus that for stator and rotor losses and, when neglecting the losses, can be written as:

$$P_{mech} \cong P_{ag} + P_r = (1 - S)P_{ag} \tag{6.23}$$

So,

$$P_r = -SP_{ag}.$$

This shows that the rotor power is a linear function of slip and may be drawn from or supplied to utility grid depending on the sign of slip, which is determined by the rotor speed with the stator synchronous rotating speed. A wind turbine driving a DFIG can have three operation modes:

(1) Super-synchronous operation, when $N_r > N_s$
As already described in Sec. 6.4.2, slip $S < 0$, the rotor current flows in the reverse direction to that in motoring mode, P_r is positive hence the generated power is delivered to the grid through both rotor and stator terminals. The total power to the network is thus:

$$P_{net} = (1 - S)P_s = P_s + P_r.$$

(2) Sub-synchronous operation, $N_r < N_s$
In this case slip $S > 0$, the rotor draws power from the grid, hence P_r is negative. Machine-generated power is delivered to the grid only through the stator, so:

$$P_{net} = (1 - S)P_s = P_s - P_r.$$

(3) Synchronous operation, when $N_r = N_s$,

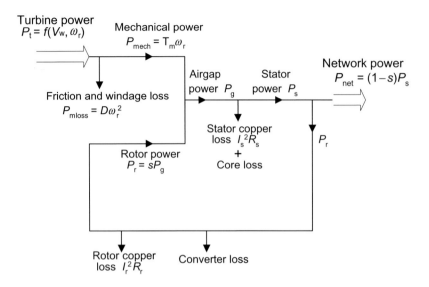

Fig. 6.24 Power flow diagram for a DFIG.

Under this condition, $S = 0$, grid supplies DC excitation current to the rotor through rotor-side power converter and only stator delivers power to the grid. Neglecting rotor and converter losses $P_{net} = (1 - S)P_s = P_s$.

Generally, for wind turbine operation, the rotor speed range is limited to approximately 33% above and below synchronous speed so the converter power rating is about a third of the stator power. The power supplied to the network can also be represented in terms of the resultant input power of the system and the total electrical losses P_{eloss} as:

$$P_{net} = P_{mech} - P_{eloss}, \qquad (6.24)$$

where P_{eloss} = stator copper loss + core loss + rotor copper loss + converter loss. A power flow diagram for a DFIG is shown in Fig. 6.24.

6.5 Analysis of Induction Generator in Terms of Complex Vector Representation[13,14]

In order to study the performance of three-phase machines, both synchronous and induction, under transient and steady-state conditions, it is necessary to develop equations that describe instantaneous time variations of the variables. The analysis is based on equivalent circuit concepts of the machines and follows a step-by-step pattern. The following assumptions are made for developing the model.

- An idealised circuit model of the machine is used in which magnetic saturation and leakages are ignored, i.e. the stator and rotor iron have infinite permeability.
- The machine has a symmetrical three-phase and star-connected stator winding with neutral point electrically isolated.
- Core losses due skin and hysteresis effects are neglected.
- Effects of V/I harmonics are neglected.

Flux linkage equations for the various self- and mutual inductances of the six windings, shown in Fig. 6.25, are established together with basic voltage — current relations.

Consider the stator-base stationary reference frame, the circuit equations for the six circuits of Fig. 6.25 may be written as:

$$v_{as} = R_s i_{as} + \frac{d}{dt}\lambda_{as} \tag{6.25}$$

$$v_{bs} = R_s i_{bs} + \frac{d}{dt}\lambda_{bs} \tag{6.26}$$

$$v_{cs} = R_s i_{cs} + \frac{d}{dt}\lambda_{cs} \tag{6.27}$$

$$v_{ar} = R_r i_{ar} + \frac{d}{dt}\lambda_{ar} \tag{6.28}$$

$$v_{br} = R_r i_{br} + \frac{d}{dt}\lambda_{br} \tag{6.29}$$

$$v_{cr} = R_r i_{cr} + \frac{d}{dt}\lambda_{cr}. \tag{6.30}$$

Corresponding flux linkage equations represent $\lambda_{as}, \lambda_{bs}, \lambda_{cs}, \lambda_{ar}, \lambda_{br}$, and λ_{cr} in terms of the circuit current $i_{as}, i_{bs}, i_{cs}, i_{ar}, i_{br}$, and i_{cr} and values of the self- and mutual inductance coefficients in E 6.31. Equations (6.25–6.30) can be

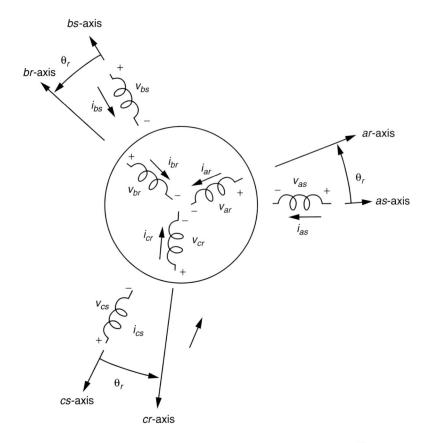

Fig. 6.25 Magnetic axes of a three phase induction machine.[13]

written in matrix form as:

$$U = Ri + pLi,$$

where

$$U = \begin{bmatrix} v_{as} \\ v_{bs} \\ v_{cs} \\ v_{ar} \\ v_{br} \\ v_{cr} \end{bmatrix} ; \quad i = \begin{bmatrix} i_{as} \\ i_{bs} \\ i_{cs} \\ i_{ar} \\ i_{br} \\ i_{cr} \end{bmatrix} ; \quad p = \frac{d}{dt}$$

$$R = \begin{vmatrix} R_s & 0 & 0 & 0 & 0 & 0 \\ 0 & R_s & 0 & 0 & 0 & 0 \\ 0 & 0 & R_s & 0 & 0 & 0 \\ 0 & 0 & 0 & R_r & 0 & 0 \\ 0 & 0 & 0 & 0 & R_r & 0 \\ 0 & 0 & 0 & 0 & 0 & R_r \end{vmatrix}$$

$$L = \begin{vmatrix} L_s & -\frac{1}{2}L_{sn} & -\frac{1}{2}L_{sm} \\ -\frac{1}{2}L_{sm} & L_s & -\frac{1}{2}L_{sm} \\ -\frac{1}{2}L_{sm} & -\frac{1}{2}L_{sm} & L_s \\ M_{sr}\cos\theta_r & M_{sr}\cos\left(\theta_r - \frac{2}{3}\pi\right) & M_{sr}\cos\left(\theta_r - \frac{4}{3}\pi\right) \\ M_{sr}\cos\left(\theta_r - \frac{2}{3}\pi\right) & M_{sr}\cos\left(\theta_r - \frac{4}{3}\pi\right) & M_{sr}\cos\theta_r \\ M_{sr}\cos\left(\theta_r + \frac{2}{3}\pi\right) & M_{sr}\cos\theta_r & M_{sr}\cos\left(\theta_r - \frac{2}{3}\pi\right) \\ M_{sr}\cos\theta_r & m_{sr}\cos\left(\theta_r - \frac{2}{3}\pi\right) & M_{sr}\cos\left(\theta_r - \frac{4}{3}\pi\right) \\ M_{sr}\cos\left(\theta_r - \frac{2}{3}\pi\right) & M_{sr}\cos\left(\theta_r - \frac{4}{3}\pi\right) & M_{sr}\cos\theta_r \\ M_{sr}\cos\left(\theta_r + \frac{2}{3}\pi\right) & M_{sr}\cos\theta_r & M_{sr}\cos\left(\theta_r - \frac{2}{3}\pi\right) \\ L_s & -\frac{1}{2}L_{sm} & -\frac{1}{2}L_{sm} \\ -\frac{1}{2}L_{sm} & L_s & -\frac{1}{2}L_{sm} \\ -\frac{1}{2}L_{sm} & -\frac{1}{2}L_{sm} & L_s \end{vmatrix},$$

(6.31)

in which L_s and L_r represent stator and rotor self-inductances respectively. When the leakage inductances are ignored, M_{sr} is the stator and rotor mutual inductance, and θ_r is the angular difference between the axes of stator and rotor windings.

As is seen, the above is an algebraically complicated set of non-linear differential equations. The analytical non-linearity is caused by time variation of the alignment angle θ_r. The three-phase machine can be more conveniently analysed in terms of complex space vectors or space phasors.

6.5.1 Three-phase to d-q-0 space vector transformation

An appropriate transformation from a-b-c variables into d-q-0 variables is found to greatly simplify the algebra. This arises from splitting the airgap mmfs for both stator and rotor along two perpendicular axes, the direct axis d and the quadrature axis q (Fig. 6.26). The common frame of reference can be stationary or rotate with the rotor or rotate with the machine synchronous speed ω.

Thus, for balanced three-phase stator currents $i_{as}, i_{bs},$ and $i_{cs},$ the complex stator current vector is defined by:

$$\bar{i}_s = i_{as}(t) + a i_{bs}(t) + a^2 i_{cs}(t) = |i_s(t)| e^{j\omega t} \qquad (6.32)$$

where operator $a = e^{j2\pi/3}$ and $a^2 = e^{j4\pi/3}$.

For balanced three-phase sinusoidal currents, the vector $\bar{i}_s$ has a constant amplitude and rotates with constant angular velocity, representing the uniformly rotating field for normal three-phase operation on a balanced sine

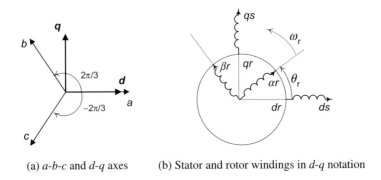

(a) a-b-c and d-q axes   (b) Stator and rotor windings in d-q notation

Fig. 6.26 Relationship between a-b-c and d-q axes and model of an induction machine.

wave supply. Equation 6.32 is a general equation that defines the stator current vector $\bar{i}_s$ at any instant and is valid only when:

$$i_{as}(t) + i_{bs}(t) + i_{cs}(t) = 0. \tag{6.33}$$

The stator current vector can be written in d-q coordinates when the d axis is placed such that it coincides with the phase a axis in the stator a-b-c model; at $t = 0$, its displacement from the phase a axis at any time t is ωt. The q axis is 90° ahead of the d axis in the direction of rotation as shown in Fig. 6.26.

Thus, we have:

$$\bar{i}_s = i_{as}(t) + a i_{bs}(t) + a^2 i_{cs}(t) = i_{ds} + j i_{qs}.$$

The a-b-c-to-d-q transformation for stator current is:

$$\begin{bmatrix} i_{ds} \\ i_{qs} \end{bmatrix} = \frac{2}{3} \begin{bmatrix} 1 & -\frac{1}{2} & -\frac{1}{2} \\ 0 & \frac{\sqrt{3}}{2} & -\frac{\sqrt{3}}{2} \end{bmatrix} \begin{bmatrix} i_{as} \\ i_{bs} \\ i_{cs} \end{bmatrix}. \tag{6.34}$$

Note Eq. (6.34) can be equally applied to three-phase voltage and flux linkage. The inverse transformation (d-q-to-a-b-c) of Eq. (6.34) is given by:

$$i_{as} = i_{ds}$$

$$i_{bs} = -\frac{1}{2} i_{ds} + \frac{\sqrt{3}}{2} i_{qs}$$

$$i_{cs} = -\frac{1}{2} i_{ds} - \frac{\sqrt{3}}{2} i_{qs}.$$

The above three-phase to two-phase transformation can be applied to three-phase rotor currents i_{ar}, i_{br}, and i_{cr}. Thus, we have:

$$\bar{i}_r = i_{ar}(t) + a i_{br}(t) + a^2 i_{cr}(t) = i_{\alpha r} + j i_{\beta r}.$$

The use of $\alpha - \beta$ notation is because $\bar{i}_r$ is defined with respect to a rotor reference frame that is coincide with the axis of the ar phase. The rotor winding also has an isolated neutral point hence $i_{ar} + i_{br} + i_{cr} = 0$.

Using the above transformation for the voltage and flux linkage, and defining voltages and currents in the stator and rotor of the machine in their natural reference frames (ds-qs, αr − βr), the machine equations can be expressed in a matrix form as:[14]

$$\begin{bmatrix} v_{ds} \\ v_{qs} \\ v_{\alpha r} \\ v_{\beta r} \end{bmatrix} = \begin{bmatrix} R_s + pL_s & 0 & pL_m \cos\theta_r & -pL_m \sin\theta_r \\ 0 & R_s + pL_s & pL_m \sin\theta_r & pL_m \cos\theta_r \\ pL_m \cos\theta_r & pL_m \sin\theta_r & R_r + pL_r & 0 \\ -pL_m \sin\theta_r & pL_m \cos\theta_r & 0 & R_r + pL_r \end{bmatrix} \begin{bmatrix} i_{ds} \\ i_{qs} \\ i_{\alpha r} \\ i_{\beta r} \end{bmatrix},$$

(6.35)

where $\theta_r = \omega_r t$. However, the differential equations in the model contain time-varying coefficients θ_r, which changes the impedance value in Eq. (6.35), according to the instantaneous rotor position.

It is possible to apply a vector rotation that transforms stator and rotor quantities to common stationary reference frame fixed to the stator d and q axes using coordinate transformations:

$$\begin{bmatrix} i_{dr} \\ i_{qr} \end{bmatrix} = \begin{bmatrix} \cos\theta_r & -\sin\theta_r \\ \sin\theta_r & \cos\theta_r \end{bmatrix} \begin{bmatrix} i_{\alpha r} \\ i_{\beta r} \end{bmatrix}.$$

(6.36)

Equation (6.35) can then be rewritten as:

$$\begin{bmatrix} v_{ds} \\ v_{qs} \\ v_{dr} \\ v_{qr} \end{bmatrix} = \begin{bmatrix} R_s + pL_s & 0 & pL_m & 0 \\ 0 & R_s + pL_s & 0 & pL_m \\ pL_m & \omega_r L_m & R_r + pL_r & \omega_r L_r \\ -\omega_r L_m & pL_m & -\omega_r L_r & R_r + pL_r \end{bmatrix} \begin{bmatrix} i_{ds} \\ i_{qs} \\ i_{dr} \\ i_{qr} \end{bmatrix}.$$

(6.37)

The impedance matrix of Eq. (6.37) contains constant coefficients which no longer depend on the rotor position. Thus, Eq. (6.37) may be used as the electrical part of an induction machine model.

The mechanical part of the generator model is defined by the relationship between developed torque and rotor speed given as:

$$T_m - T_e = Jp\omega_r + D\omega_r,$$ (6.38)

where T_m is the turbine torque and T_e is the generator torque and is given by:

$$T_e = \frac{3}{2} ppL_m(i_{qs}i_{dr} - i_{ds}i_{qr}).$$ (6.39)

Equation (6.37) to (6.39) define a complete mathematical model of an induction generator in the stator stationary reference frame.

6.6 Switched Reluctance Machines

The switched reluctance drive (SRD) is a development of great promise, but has so far achieved only small penetration into the rotating electrical drives market. The machines are alternatively known as variable reluctance motors, a term that reflects the technology.

6.6.1 *Switched reluctance motor*[8,9]

Motoring action is a high-power development of the single-stack and variable reluctance stepper motor. An example (Fig. 6.27) has eight stator poles and six rotor teeth. The doubly salient structure generally has a different number of poles on the rotor and the stator. There are no rotor windings

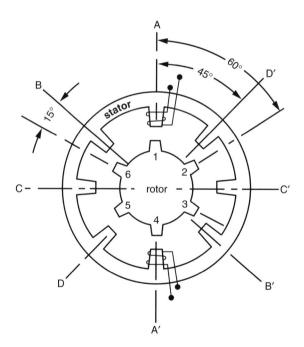

Fig. 6.27 Single-stack, four-phase, eight/six-pole variable reluctance steeper motor (only the phase A winding is shown).

and short stator-end windings that facilitate easy cooling. The concentrated stator windings are energised by a pulse train resulting in a continuous rotational speed that is synchronous to the pulse frequency. The energising pulses are provided by a power electronics-switching package that is custom designed to match each particular motor. Very high speeds can be realised and the high torque–inertia ratio makes it ideal as servomotor. The motor has zero open-circuit voltage and contributes zero current to an external fault. Its efficiency is comparable with that of the equivalent induction motor. Closed-loop control via rotor positional feedback permits very precise position control.

A switched reluctance motor torque is due entirely to reluctance variation developed by the saliencies of one or two pairs of stator and rotor poles. This is in contrast with the synchronous reluctance motor, with which it has several important properties in common, where the entire airgap surface contributes to the energy conversion process. Both the switched reluctance drive and the synchronous reluctance motor have in common that they require only unipolar current energisation since the rotor is not polarised. This reduces the number of switching devices required in the electronic controller package. Sometimes bipolar drive circuits are used. It is necessary to recover energy from the excited windings at current switch-off, which can be realised by the use of bifilar winding on the stator poles.

A switched reluctance motor drive can provide four-quadrant torque-speed operation at 200 kW. Careful hardware design and complex algorithms are needed to address the problems of torque pulsations and acoustic noise. The very high level of controllability of a switched reluctance motor makes it a natural competitor to the brushless DC motor drive and the vector-controlled induction motor in adjustable speed drives.

6.6.2 *Switched reluctance generator*[10,11]

The switched reluctance generator, like the switched reluctance motor, is a doubly salient machine with no magnet or brushes. As the iron rotor poles, are driven past the stator poles, there is a changing reluctance of generator magnetic circuit. This creates changes in the inductance of the stator poles that causes a current to be induced in the stator windings. A pulse current waveform then appears at each stator pole.

In polyphase machines, the outputs from each phase are fed to a converter which switches each phase sequentially to a DC link to provide a direct voltage. There needs to be position sensing on the rotor shaft to control the trigger timing of the converter switches. The sensors enable the current to be controlled by varying the turn-on and turn-off angles of the output current, depending on the pole position.

Switch reluctance generators feature compact and robust designs that are highly suited to variable speed operation, sometimes at very high speeds in (say) aircraft applications. The machine is relatively inexpensive to manufacture but has received little attention, except in industrial feasibility studies, for the high-power slow-speed operation of wind generators. There is no inherent reason why switched reluctance generator machines should not be used.

6.7 What Form of Generator is the Best Choice for Wind Generation Systems?

Both synchronous machines and induction machines have been and continue to be extensively used as generators in wind-energy generation schemes. Neither form of machine is intrinsically superior or inferior to the other form. Each form of machine has properties that are appropriate in different circumstances.

The cyclic torque delivered by a wind turbine to the generator contains torsional oscillations. In analytical terms, the oscillations can be thought of as containing higher harmonics of torque. The main cyclic torque is at blade passing frequency, referred to in Sec. 2.5.3 of Chap. 2, which sometimes matches the natural frequency of oscillation of a small synchronous generator connected to an electrical network.

The multi-megawatt synchronous generators used almost universally, in electrical power plant generation systems, are designed with cage damper windings to suppress torque harmonics. But, it is not practical to provide embedded damper windings in the much smaller synchronous generators rated for wind-energy applications. If the power train ("drive" train) of a wind turbine system contains a synchronous generator, then there is no built-in damping to control the torque oscillations and other methods have to be used.

An induction generator acts intrinsically as a torsional damper mechanism, where the torque is proportional to the slip speed. Certain second-order effects tend to reduce the damping available for machines above about 1 MW rating. Nevertheless, the intrinsic damping provided by induction machines is valuable.

References

1. de Vries, E., Thinking bigger, *Renewable Energy World*, **B**(3) 42–55, 2005.
2. Fitzgerald, A. E., C. Kingsley, Jr, *Electrical Machinery*, Second edition, McGraw-Hill Book Company, Inc., the USA, 1961.
3. Gray, C. B., *Electrical Machines and Drive Systems*, Longman Scientific and Technical, Harlow, Essex, England.
4. Alger, P. L., *Induction Machines*, Second Edition, Gordon and Breach Science Publishers, New York, the USA, 1970.
5. Say, M. G., *Alternating Current Machines*, Fifth Edition, Pitman Publishing Ltd., London, England, 1983.
6. Mcpherson, G., *An Introduction to Electrical Machines and Transformers*, Second Edition, John Wiley & Sons, New York, the USA, 1990.
7. Burton, T., D. Sharpe, N. Jenkins, E. Bossanyi, *Wind Energy Handbook*, John Wiley & Sons Ltd., Chichester, England, 2001.
8. Shepherd, W., L. N. Hulley, D. T. W. Liang, *Power Electronics and Motor Drives*, Second Edition, Cambridge University Press, Cambridge, England, 1998.
9. Murphy, J. M. D., F. G. Turnbull, *Power Electronic Control of AC motors*, Pergamon Press, Oxford, England, 1988.
10. Gardner, P., *Electrical Issues Associated with Variable Speed Operation of Wind Turbines*, Contract Report ETSU W/33/ 00417/REP.
11. Power, M., *Electric Drives — Generators*, http://www.mpoweruk.com/generators.htm, March 2008.
12. Slootweg, H., E. de Vries, Inside wind turbines (fixed and variable speed), *Renewable Energy World*, **6**(1) 31–40, 2003.
13. Novotny, D. W., T. A. Lipo, *Vector Control and Dynamics of AC Drives*, Oxford Science Publication, Oxford University Press, New York, the USA, 1996.
14. Vas, P., *Vector Control of AC Machines*, Oxford Science Publishing, 1990.

CHAPTER 7

Power Electronic Converters in Wind-Energy Systems

7.1 Types of Semiconductor Switching Converters

Semiconductor switching converters may be grouped into four main categories, according to their functions.[1]

1. The transfer of power from an alternating current (AC) supply to a direct current (DC) form. This type of converter is usually called a *rectifier*.
2. The transfer of power from a DC supply to an AC form. This type of converter is usually called an *inverter*.
3. The transfer of power from an AC supply directly into an AC load of different frequency. This type of converter is called a *cycloconverter* or a *matrix converter*.
4. The transfer of power from a DC supply directly into a DC load of different voltage level. This type of converter is called a *chopper converter* or a *switch-mode converter*.

The DC to DC type of converter, item 4, is not normally used in wind-energy conversion systems.

The general case requirement is that a set of three-phase voltages, variable in magnitude and frequency, requires to be converted to a set of three-phase voltages, of firmly fixed values of magnitude and frequency,

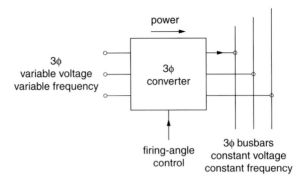

Fig. 7.1 Basic form of converter in wind-energy systems.

specified by the rating of the three-phase power system into which it is to be synchronised, Fig. 7.1.

There exist classes of converters known as cycloconverters and matrix converters, which perform direct AC to AC conversion using power electronic switching.[2] These are not commonly used for wind-energy applications. The form of converter most usually employed is the three-phase, full-wave converter, in two stages, illustrated in Fig. 7.2. A controlled rectifier converts the input voltages to a direct voltage at the DC link. This is then used as the input to an inverter stage that transfers the power into the three-phase infinite bus bars of the power system, via a transformer if required.

Both the rectifier stage and the inverter stage use the same basic form of circuitry, which is the three-phase bridge connection. Power flow through the bridges is controlled by the adjustment of the firing-angles α of the bridge switches in the case of silicon-controlled rectifier (SCR) switches or by the adjustment of the gate (or base) currents in the cases of power transistor switches.

7.2 Three-Phase Controlled Bridge Rectifier[1–3]

Wind-energy rotating generators, both synchronous and asynchronous (induction) types, usually have three-phase and star-connected (wye-connected) armature windings that form the input to the bridge rectifier stage (Fig. 7.3).

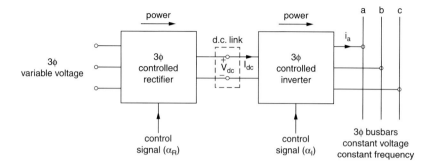

Fig. 7.2 Basic arrangement of the three-phase bridge rectifier–inverter.

When any electrical supply system has a low (ideally zero) impedance, the supply voltages remain sinusoidal at all the times. With a bridge rectifier in a wind-energy system, the rectifier switching action causes non-sinusoidal pulses of current to be drawn from the generator. The supply voltages at the bridge input terminals are slightly distorted but can usually be regarded as sinusoidal. For the purpose of circuit analysis, one can assume that semiconductor rectifier elements such as diodes and silicon-controlled rectifiers are ideal switches. During conduction they are dissipationless and have zero voltage drop. Also, when held in extinction by reverse anode voltage, they have infinite impedance.

Although the switching action of the bridge rectifier can create some significant problems in the analysis of the wind generator itself, these do not cause any difficulty in the steady-state analysis of the rectifier and its load.

The voltages at the bridge input terminals a, b, and c can be defined in terms if the angular frequency ω of the wind generator by:

$$e_{aN} = E_m \sin \omega t \tag{7.1}$$

$$e_{bN} = E_m \sin(\omega t - 120°) \tag{7.2}$$

$$e_{cN} = E_m \sin(\omega t - 240°). \tag{7.3}$$

The corresponding line-to-line voltages at the supply point are:

$$e_{ab} = e_{aN} + e_{Nb} = e_{aN} - e_{bN} = \sqrt{3} E_m \sin(\omega t + 30°) \tag{7.4}$$

$$e_{bc} = \sqrt{3} E_m \sin(\omega t - 90°) \tag{7.5}$$

$$e_{ca} = \sqrt{3} E_m \sin(\omega t - 210°). \tag{7.6}$$

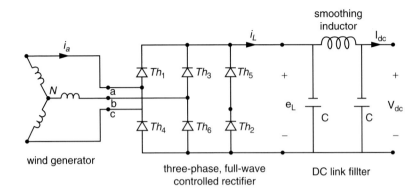

Fig. 7.3 Wind generator feeding the rectifier stage.

Time variations of the two forms of the input voltages are shown in Fig. 7.4 together with the output voltage $e_L(\omega t)$ at three different values of firing angle α_R. The terminology α_R is used to describe the rectifier firing angle, which is distinct from the corresponding term α_I for the inverter.

The device numbering notation shown in the bridge rectifier circuit of Fig. 7.3 is standard for the three-phase and full-wave controlled bridge in both the rectifier and the inverter modes of operation. To provide a current path from the supply side to the load side requires the simultaneous conduction of at least two appropriate switches. When one element of the upper group switches and one of the lower group conducts, the corresponding line-to-line voltage is applied directly to the load. In Fig. 7.3, the switches are depicted as the silicon-controlled rectifier type of thyristor. For this reason, the terminology Th is used in their description. If, for example, the switches Th_1 and Th_6 conduct simultaneously, then line voltage e_{ab} is applied across the load. There are some switch combinations that are not permissible. If, for example, the switches in any leg conduct simultaneously from the top half and the bottom half of the bridge, then this would represent a short circuit on the AC supply. To provide load current of the maximum possible continuity and smoothness, appropriate bridge switches must conduct in pairs sequentially, for conduction intervals up to 120° or $\pi/3$ radius of the supply voltage. The average load voltage and current are controlled by the firing angle of the bridge thyristors, each measured from the crossover point of its respective phase voltages.

When the bridge rectifier firing angle α_R is zero degrees, it then behaves like an uncontrolled (diode) rectifier. The rectifier output voltage $e_L(\omega t)$, in Fig. 7.3, has a peak value variable in magnitude dependent on the level of the wind-generator voltages. Also, its average value E_{av} is controlled by the value of the bridge rectifier firing angle α_R. Waveform $e_L(\omega t)$, shown in Fig. 7.4, has a dominant DC value but also contains pulsations of sixth harmonic wind-generator frequency. A typical section of the waveform $e_L(\omega t)$, in the interval $\alpha_R + 30° \leq \omega t \leq \alpha_R + 90°$, is described by:

$$e_L(\omega t) = \sqrt{3} E_m \sin(\omega t + 30°)|_{\alpha+30°}^{\alpha+90°}. \tag{7.7}$$

The average value E_{av} of Eq. (7.7) in terms of peak phase voltage E_m is:

$$\begin{aligned} E_{av} &= \frac{3}{\pi} \int_{\alpha_R+30°}^{\alpha_R+90°} \sqrt{3} E_m \sin(\omega t + 30°) d\omega t \\ &= \frac{3\sqrt{3}}{\pi} E_m \cos \alpha_R \\ &= E_{av_0} \cos \alpha_R, \end{aligned} \tag{7.8}$$

where

$$E_{av_0} = \frac{3\sqrt{3}}{\pi} E_m = 1.654 E_m. \tag{7.9}$$

Consistent with Eq. (7.8), the average voltage level E_{av} is shown in Fig. 7.4 as being proportional to $\cos \alpha_R$. In Fig. 7.3(e), for example, E_{av} is one-half or $\cos 60°$ times the value given in Fig. 7.3(c) when $\alpha = 0°$ and $\cos 0° = 1$. The average output voltage E_{av_R} can also be expressed in terms of the directly measurable rms value E of the input phase voltage since:

$$E = \frac{E_m}{\sqrt{2}}. \tag{7.10}$$

Combining Eq. (7.8) and Eq. (7.10) gives:

$$\begin{aligned} E_{av_R} &= \frac{3\sqrt{3}}{\pi} \sqrt{2} E \cos \alpha_R \\ &= 2.34 E \cos \alpha_R. \end{aligned} \tag{7.11}$$

Expression (7.11) provides a means of calculating the thyristor firing angle α_R in terms of readily measurable electrical voltages.

The waveform of the current $i_L(\omega t)$ leaving a bridge rectifier, is determined largely by the electrical nature of the connected load. In this case, the connected load usually consists of the DC link filter in cascade with a three-phase and controlled inverter feeding into an infinite bus, as shown in Fig. 7.1. Seen from the generator, the load appears to be a large inductance. This causes the currents in the generator windings and in the bridge switches to be constant amplitude pulses. The bridge output current $i_L(\omega t)$ is very close to being ideal DC, of value I_{dc}, as shown in Fig. 7.5. Adjustment of the bridge firing angle α_R causes the amplitude I_{dc} to change in accordance with the change of voltage E_{av} from Eq. (7.11).

The most general form of DC link filter (Fig. 7.6) contains a series inductor having resistance R_f and self-inductance L_f. Its function is to smooth

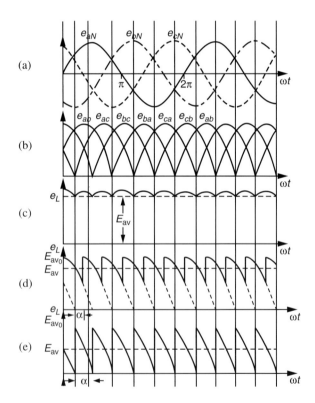

Fig. 7.4 Voltage waveforms of the three-phase controlled bridge rectifier (a) input phase voltages, (b) input line-to-line voltages, (c) output voltage $e_L(\omega t)$ at $\alpha_R = 0°$, (d) output voltage at $\alpha_R = 30°$, (e) output voltage at $\alpha_R = 60°$.

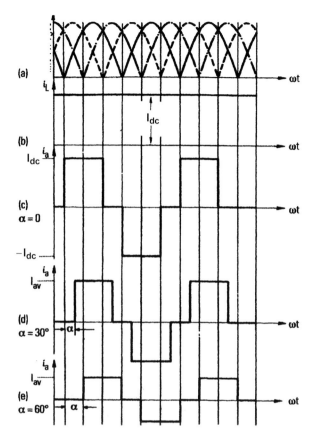

Fig. 7.5 Waveforms of the three-phase, full-wave controlled bridge rectifier circuit highly inductive load and ideal supply: (a) supply line voltages, (b) load current ($\alpha_R = 0°$), (c) supply line current $i_a(\alpha_R = 0°)$, (d) supply line current $i_a(\alpha_R = 30°)$, and (e) supply line current $i_a(\alpha_R = 60°)$.

out perturbations of the link current so that the output current consists primarily of a direct component I_{dc}. The two shunt-connected capacitors C, in Fig. 7.6, do not draw any significant amount of current and are included to smooth out perturbations in the bridge output voltage, which can be regarded as a purely direct voltage V_{dc}. In reality, both the output voltage V_{dc} and the output current I_{dc} contain small time-varying components but these can usually be ignored for calculation purposes. Some forms of filter contain only the series inductor without any capacitance and some contain the inductor plus one capacitor

Since the inductor contains resistance R_f, it dissipates a power $I_{dc}^2 R_f$ as heat. This power has to be supplied from the wind generator through the rectifier bridge and represents a loss of efficiency in the system. Since the current in the inductor is virtually a pure DC, it causes no voltage drop due to the inductance L_f.

Electrical power can be transmitted only by the product of voltage and current components of the same frequency. At the DC link in Fig. 7.2, the output voltage consists largely of a DC component defined by Eq. (7.8). After filtering, this becomes the direct voltage V_{dc}. Now V_{dc} is, by definition, of zero frequency. This voltage can transmit power into the next stages, the DC link and the inverter, only by combining with a zero-frequency component of current. This component is the DC link current I_{dc}. The power passing out of the DC link is (very nearly):

$$P_{link} = V_{dc} I_{dc}. \tag{7.12}$$

The real power P_{link} becomes zero if $E_{av} = 0$ (making $V_{dc} = 0$) or $I_{dc} = 0$ or both. In Eq. (7.11), the switch condition to make $E_{av} = 0$ is that $\alpha_R = 90°$ or that the input voltage from the wind generator is zero.

The power out of the DC link filter is equal to the power output from the wind generator minus the relatively small power losses in the link filter components and the switching losses of the bridge rectifier switching devices.

7.3 Three-Phase Controlled Bridge Inverter Feeding an Infinite Bus

7.3.1 *Output voltage*

The basic form of a three-phase and controlled bridge inverter circuit is shown in Fig. 7.7. For inverter operation, the output voltage level, frequency and waveform, are set by the AC bus and cannot be changed. In an inverter circuit, the configuration and notation of the switching devices are exactly the same as for a rectifier circuit. As with rectifier operation, described in the preceding section, the anode voltages of the switches undergo cyclic variation and are therefore switched off by natural commutation. There is no advantage to be gained here by the use of gate turn-off devices.

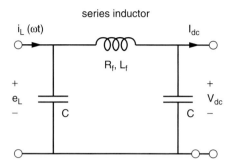

Fig. 7.6 DC link filter.

For the circuit of Fig. 7.3, certain restrictions must be imposed on the switching sequences, similar to those described in the preceding subsection.

If sequential firing is applied to the six bridge SCRs the three-phase currents are identical in form but mutually displaced in phase by 120°. The detailed operation of the inverter circuit is similar to that described for rectifier operation in the preceding section, except that the polarity of the direct voltage V_{dc} is reversed and the range of firing angles is now $90° \leq \alpha_I \leq 180°$, where the terminology α_I is now used for the inverter firing angle.

The current $i_a(\omega t)$ injected into the three-phase system (Fig. 7.7), can be idealised as having a waveform of rectangular pulses, each of wavelength

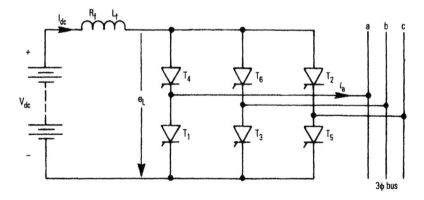

Fig. 7.7 Three phase, controlled bridge inverter feeding into an infinite bus system.[1]

120° of the bus angular frequency, similar to the rectifier waveform of Fig. 7.5. The maximum value of the current pulses is equal to the current I_{dc} at the inverter input which, in turn, is dependent on the rectifier firing angle α_R.

All electrical supply systems contain small amounts of series inductance. This has the effect of slightly delaying the onset of conduction in a circuit after switching occurs. The phenomenon is known as "overlap" and is demonstrated in Fig. 7.8(b). Overlap has no effect on the peak value I_{dc} of the phase current and is not considered in the present analysis. Idealised current waveforms are used in Fig. 7.9 to demonstrate the effect of firing-angle control in the range of $90° \leq \alpha_I \leq 150°$.

The average value of the inverter output voltage is related to the firing angle α_I and the electrical output parameters by:

$$E_{av_I} = \frac{3\sqrt{3}}{\pi} E_{mb} \cos \alpha_I, \qquad (7.13)$$

where E_{mb} is the peak phase voltage of the AC bus. It is seen that Eq. (7.13) is precisely the same form as Eq. (7.8) for the rectifier. In equivalent circuit terms, the rectifier and inverter combination can be represented, in idealised form, by Fig. 7.10.

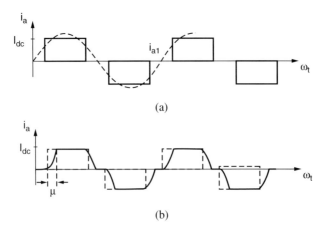

Fig. 7.8 Phase current $i_a(\omega t)$ for the three-phase inverter at $\alpha = 0°$ (a) with a perfect (impedanceless) three-phase system (b) exhibiting overlap due to the presence of three-phase system series inductance.

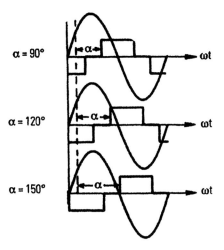

Fig. 7.9 Instantaneous phase voltage $e_a(\omega t)$ and current $i_a(\omega t)$ on the AC side of a bridge inverter, assuming ideal AC supply (*Note*: It is assumed that V_{dc} is adjusted proportionately to $\cos \alpha_I$ to maintain I_{dc} constant).

The source of power is the wind generator plus the controlled rectifier that creates an adjustable voltage V_{dc} at the DC link is given by Eq. (7.8). Power is injected into the AC bus, via the inverter, by voltage V_{dc} acting as a driving voltage, given by Eq. (7.13). It should be noted that Fig. 7.9 has been simplified by the omission from the equations of the effects of overlap in both the rectifier and (separately) the inverter.

For the calculation of power output from the inverter into the AC system, it has to be noted that the frequency is, by definition, the electrical system frequency. Real power in watts is transmitted from the inverter into the system only by combining the output voltage with the component of current that is also of system frequency. In terms of Fourier analysis, the sinusoidal phase voltage of the AC system has to be combined, with the fundamental (i.e., system frequency) component of the rectangular pulses current wave. A sketch of typical fundamental current Fourier harmonic component i_{a_1} is shown in Fig. 7.8(a). It is important to note that the Fourier analysis and its related concept of harmonic components are entirely theoretical. The fundamental Fourier component, like all the other harmonic components, has no physical existence. It cannot be seen on a cathode ray oscilloscope because there is nothing physical to see.

Nevertheless, the concept of dividing up a complex wave into a series of sinusoidal harmonics is a brilliant idea. It is a very useful theoretical concept in circuit analysis and sometimes offers insight into the physical operation of the circuit.

7.3.2 Real (average) power output

The power entering the electrical system is the inverter output power. For any three-phase system, the total real or active power P, for all three phases, is given by:

$P = 3 \times$ phase voltage

$\times$ rms value of the phase current component of the same frequency as the voltage

$\times$ cosine of the angle between the voltage and the current

$$= 3 \frac{E_{mb}}{\sqrt{2}} I_{a_1} \cos < \frac{\frac{E_{mb}}{\sqrt{2}}}{I_{a_1}}$$

$$= 3 \frac{E_{mb}}{\sqrt{2}} I_{a_1} \cos \psi_1. \qquad (7.14)$$

Now for the three-phase bridge circuits, it can be shown that:[1,2]

$$\psi_1 = \alpha_I. \qquad (7.15)$$

The displacement factor $\cos \psi_1$ of the output power is therefore equal to the cosine of the inverter thyristor firing angle:

$$\cos \psi_1 = \cos \alpha_I. \qquad (7.16)$$

Combining Eq. (7.12) with Eq. (7.16) gives a useful expression for the real power delivered into the power network:

$$P = 3 \frac{E_{mb}}{\sqrt{2}} I_{a_1} \cos \alpha_1$$

$$= 3 E_b I_{a_1} \cos \alpha_1, \qquad (7.17)$$

where E_b is the per-phase value of the bus voltage. It should be noted that although the relationship of Eq. (7.17) is valid here for the case of an inverter feeding into a power network, it is not true for the more basic isolated circuit of a bridge rectifier with resistive loading.

The rms value of the fundamental component of current, I_{a_1}, for the current wave shape of Figs. 7.8 and 7.9 is numerically related to the peak input current I_{dc} by:

$$I_{a_1} = \sqrt{2}\frac{\sqrt{3}}{\pi}I_{dc} = \frac{\sqrt{6}}{\pi}I_{dc} = 0.78 I_{dc}. \quad (7.18)$$

This fundamental component of current therefore has a peak value:

$$\hat{I}_{a_1} = \sqrt{2}I_{a_1} = \sqrt{2}\,0.78\,I_{dc} = 1.1\,I_{dc}. \quad (7.19)$$

The fundamental current $i_{a_1}(\omega t)$ sketched in Fig. 7.8(a) has a peak value $\sqrt{2}I_{a_1}$ and is in time phase with the current $i_a(\omega t)$. If the small power loss in the inverter switches is neglected the inverter output power in Eq. (7.14) is equal to the input power in Eq. (7.12). Incorporating the identity of Eq. (7.16), it is seen that:

$$V_{dc} I_{dc} = 3\frac{E_{mb}}{\sqrt{2}}I_{a_1}\cos\alpha_I. \quad (7.20)$$

Combining the RHS of Eq. (7.20) with Eq. (7.18) eliminates I_{dc} to result in:

$$V_{dc} = 3\frac{E_{mb}}{\sqrt{2}}0.78\cos\alpha_I$$

$$= 1.65 E_{mb}\cos\alpha_I. \quad (7.21)$$

Since the value of $E_{mb}(=\sqrt{2}E_b)$ is prescribed by the electrical power system, there is a direct relationship in Eq. (7.21) between the DC link voltage output V_{dc} and the corresponding inverter firing angle α_I. The rectifier inverter power link is illustrated by the equivalent circuit of Fig. 7.10.

In order to increase the power input P into the AC system, assuming constant system voltage E_b, it is seen from Eq. (7.17) that it is necessary to increase either the value of the injected current I_{a_1} (and therefore I_a) or the value of $\cos\alpha_I$ or both. Also, from Eq. (7.19), in order to increase I_{a_1}, it is necessary to increase the DC link current I_{dc}.

7.3.3 Reactive power

The operation of the three-phase bridge inverter incurs a reactive power requirement Q. Although the real power, P, transferred through the inverter comes from the DC source, the reactive power cannot be transferred through the DC link. The reactive voltamperes have to be provided by the AC system or by parallel connected capacitance at the inverter output terminals.

The reactive power Q passing between the inverter and the AC power system is given by an equation that is the complement of Eq. (7.17):

$$Q = 3E_b I_{a_1} \sin \alpha_1. \tag{7.22}$$

The total apparent power S in a sinusoidal system was shown in Eq. (5.25) of Chap. 5 to be:

$$S = \sqrt{P^2 + Q^2}. \tag{7.23}$$

Using P from Eq. (7.17) and Q from Eq. (7.22) in Eq. (7.23)

$$P + Q = 3EI_{a_1}\sqrt{\cos^2 \alpha_I + \sin^2 \alpha_I}$$
$$= 3EI_{a_1}. \tag{7.24}$$

But this is not the apparent power S at the inverter output terminals. For any AC system, the apparent power is the product of the rms current and the rms voltage. In the case of Fig. 7.7:

$$S = 3EI_a. \tag{7.25}$$

The AC-side current can be thought of as a fundamental frequency component lagging its corresponding phase voltage by $\psi_1 (= \alpha_I)$ radians plus a series of higher odd harmonics. In terms of the fundamental frequency component, the inverter action can be interpreted as drawing lagging current from the AC system although it is actually delivering leading current into the AC system.

The AC-side components, operating at supply frequency, may be represented in phasor form as shown in Fig. 7.11. This is not strictly valid because phasor diagram representation is based on the Argand diagram and applies only to sinusoidal variables. Although the current waveform (Fig. 7.8), is not sinusoidal, it has a predominant component that is the sinusoidal fundamental (supply frequency) component. In Fig. 7.11, it should be noted that $\cos(\pi - \psi_1) = -\cos \psi_1$.

7.3.4 RMS output current

Heating in the transmission system conductors and equipment due to the inverter output current depends on the total rms value I_a of the output current (not only on the fundamental component I_{a_1}).

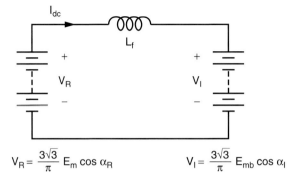

Fig. 7.10 Equivalent circuit of the rectifier-inverter link.

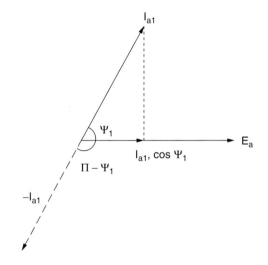

Fig. 7.11 Phasor diagram of inverter operation, $\psi_1 = 60°$.

Now, the rms value of any function $i_a(\omega t)$ that is repetitive in 2π radians is defined as:

$$I_a = \sqrt{\frac{1}{2\pi} \int_0^{2\pi} i_a^2(\omega t) \, d\omega t}. \tag{7.25}$$

Since the rms value of a function is not sensitive to the polarity of the waveform, the positive and negative parts of the waveforms of Figs. 7.8 and 7.9

have identical rms values. In calculations, it is necessary to consider only the positive (say) part of the wave and to double its value. Also, the rms value of any half wavelength in Fig. 7.8(a) is the same as the rms value of the whole wave $i_a(\omega t)$. It can be shown that for the idealised rectangular waveform of Fig. 7.8(a), the rms value I_a is related to the peak value I_{dc} by:

$$I_a = \sqrt{\frac{2}{3}} I_{dc} = 0.82 \, I_{dc}. \tag{7.26}$$

The actual apparent power given by $S = EI_a$ is slightly larger than the product EI_{a_1} using only fundamental frequency terms. The magnitude of the difference is small and can be neglected, to a first approximation.

7.3.5 Inverter power factor

The power factor of a circuit or system was defined in Eq. (5.21) of Chap. 5. For any three-phase system, the power factor is represented in Eq. (5.22). In the terminology of this chapter, the power factor (PF) of the incoming inverter in Fig. 7.7 can be written:

$$\text{PF} = \frac{P}{3E_a I_a}, \tag{7.27}$$

where P is the total three-phase power, E_a is the rms voltage per phase, and I_a is the rms current per phase.

For systems with sinusoidal terminal voltage, as in the present case, it is sometimes convenient to separate the power factor into two analytical components known as the distortion factor and the displacement factor:

$$\text{Power factor} = [\text{displacement factor}][\text{distortion factor}]. \tag{7.28}$$

The distortion factor is an indication of the extent to which the waveform of current departs from purely sinusoidal. It is defined as the ratio of the rms value of the fundamental component I_{a_1} to the rms value of the total current I_a.

$$\text{Distortion factor} = \frac{I_{a_1}}{I_a}. \tag{7.29}$$

It can be inferred that since the current waveform $i_a(\omega t)$ varies slightly with the extent of overlap (Fig. 7.8(b)), then the distortion factor also varies slightly in value, depending on the overlap. A value that is satisfactory for most purposes is obtained by using the waveform of ideal rectangular pulses of current, as shown in Fig. 7.8(a). If the value of I_{a_1} from Eq. (7.18) and the value I_a from Eq. (7.26) are substituted into Eq. (7.29), it is found that

$$\text{Distortion factor} = \frac{I_{a_1}}{I_a} = \frac{0.78}{0.82} = 0.95. \tag{7.30}$$

The same value can be obtained by an alternative analysis when it is found that:

$$\text{Distortion factor} = \frac{3}{\pi} = 0.95. \tag{7.31}$$

The distortion factor here is concerned entirely with current wave shape. It is not affected at all by reactance effects or current delay, except for the current overlap effect that has been neglected.

The displacement factor is the cosine of the fundamental phase angle ψ_1 between the terminal voltage and the fundamental component of the phase current I_{a_1}:

$$\text{Displacement factor} = \cos \psi_1 = \cos \angle \frac{E_a}{I_{a_1}}. \tag{7.32}$$

It is seen that the expression (7.32) forms part of the earlier Eq. (7.14). And since, from Eq. (7.16), the displacement angle ψ_1 is equal to the inverter firing angle α_I, then:

$$\text{Displacement factor} = \cos \alpha_I. \tag{7.33}$$

For an inverter the firing angle α_I is greater than 90° so that the value of $\cos \alpha_I$ is negative. This does not imply a negative power factor, which is meaningless, but that the current is emerging from the inverter, rather than entering it as in the case of rectifier operation.

Incorporating Eqs. (7.29) and (7.33) into Eq. (7.28) gives:

$$PF = \frac{3}{\pi} \cos \psi_1 = \frac{3}{\pi} \cos \alpha_I. \tag{7.34}$$

The PF of an inverter connected to a three-phase system is a leading PF varying from zero at $\alpha_I = 90°$ and increasing towards its maximum possible value of unity as $\alpha_I \to 180°$.

7.4 The Effect of AC System Reactance on Inverter Operation

The presence of AC system inductance in Fig. 7.7 inhibits the process of current commutation from one thyristor switch to another. Instead of the instantaneous current transfer that results in the idealised waveforms of Fig. 7.8(a) when the supply is ideal, a short but finite time is required to accomplish a complete transfer of current. The duration of the current transfer time, usually called the overlap period, depends on the magnitude of the AC system voltage, the DC current level from the DC link, the thyristor firing angle α_I, and the magnitude of the supply inductance. In Fig. 7.8(b), the overlap period is designated by the symbol μ. A fuller analysis of inverter operation in the presence of AC system inductance, given in Ref. 3, shows that there is consequent slight modification of the inverter equations developed above for ideal supply conditions.

Analytical modifications to the inverter operation due to the system reactance are summarised below.

1. The average voltage V_{dc} Eq. (7.13), is reduced.
2. The average current I_{dc} is affected by the reduction of the average voltage V_{dc} and is also reduced.
3. Because I_{dc} is reduced and V_{dc} is reduced the power in the DC link entering the inverter is also reduced, Eq. (7.12).
4. The reduction of the input power results in the reduction of the inverter rms output current I_a and also of its fundamental frequency component I_{a_1}, Eq. (7.14).
5. The reduction of V_{dc} results in the reduction of $\cos \alpha_I$, Eq. (7.21). In the firing angle range $90° \leq \alpha_I \leq 180°$, this corresponds to a reduction in value of the firing angle α_I.
6. The waveforms of the injected line currents are modified as in Fig. 7.8. This causes changes in the magnitudes of both the fundamental (AC system) frequency component and of the higher harmonic components.
7. The modification of the output current wave shape causes change of the current displacement angle Ψ_1 and consequently of the displacement factor $\cos \Psi_1$.
8. The power factor of a three-phase bridge inverter is reduced due to the overlap at the typically small values of overlap angle μ.

7.5 Three-Phase Cycloconverter Feeding an Infinite Bus

The basic conversion process is shown in Fig. 7.12. A set of three-phase, variable voltage, and variable frequency voltages is converted to a set of three-phase, constant voltage, constant frequency voltages. The output voltage, and frequency levels are fixed by the three-phase system into which power is delivered.

In a cycloconverter, there is no intermediate DC stage, as in the rectifier–inverter system of Fig. 7.2. DC to AC conversion can be realised by the use of multiple power electronic switches, usually 12 switches per phase, as shown in Fig. 7.12. Each load (bus side) phase current is shared equally between two switch groups resulting in symmetrical three-phase operation. An alternative connection (not shown) uses inter-group transformer connections to facilitate circulating currents.

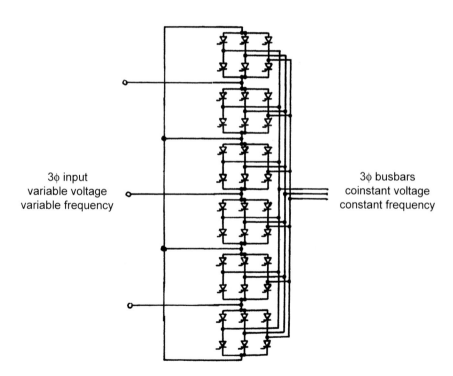

Fig. 7.12 Three-phase to three-phase cycloconverter.[2]

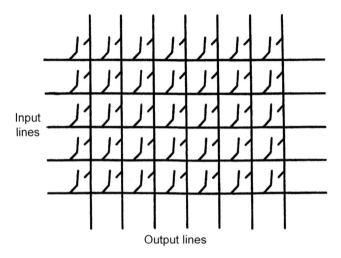

Fig. 7.13 General arrangement of an ideal switching matrix.[2]

With no DC link required, there is a significant reduction in the volume of the equipment but savings made on the cost of reactive energy storage components are offset by the greater cost of the switches and their electronic gating circuits — 36 switches in Fig. 7.12 compared with 12 (by implication) in Fig. 7.2. As with all line-commutated converters, the efficiency is high but control is achieved at the expense of having to operate with a varying lagging inverter power factor.[4] Cycloconverters have been used as the rotor converter in the induction motor slip energy recovery schemes but there the inherent low power factor of the cycloconverter exacerbates the inherent low power factor of the induction machine.

The cycloconverter has not been included, thus far, in any large-scale industrial wind-energy schemes.

7.6 Matrix Converter Feeding an Infinite Bus

The multiple conversion stages and energy-storage components of the inverter and cycloconverter circuits can, theoretically, be replaced by a switching matrix converter. An arbitrary number of input lines can be connected to an arbitrary number of output lines, using bidirectional semiconductor switches, as shown in Fig. 7.13. For the case of three-phase to three-phase conversion, required in wind-energy schemes, the basic form of

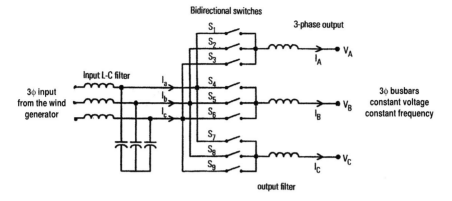

Fig. 7.14 Basic circuit arrangement for a three-phase matrix converter.[2]

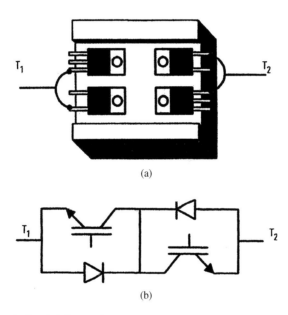

Fig. 7.15 Practical switch for matrix converter operation: (a) heat sink mounting (b) equivalent circuit.[2]

circuit is shown in Fig. 7.14. Each of the nine switches could have the configuration of Fig. 7.15, permitting blocking or conduction in either direction. The input side of the converter is a voltage source and the output is a current source. Low pass filters are needed at both the input and the output terminals

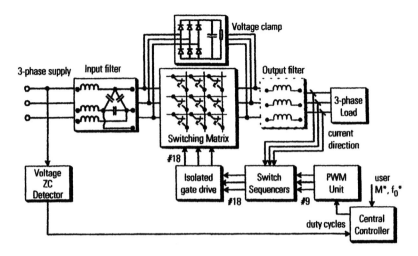

Fig. 7.16 Basic building blocks of a matrix converter.[2]

to filter out the high-frequency ripple. An overall block diagram of an experimental matrix converter system is given in Fig. 7.16. Nine pulse-width modulated (PWM) signals, generated within a programmable controller, are fed to switch sequencer circuits via pairs of differential line driver receivers. In the switch sequencers, the PWM signals are logically combined with current direction signals to produce 18 gating signals. Isolated gate driver circuits then convert the gating signals to appropriate driver signals capable of turning the power switches on or off. Each power circuit is protected by a voltage clamp circuit. A zero-crossing (ZC) detector is used to synchronise to the input voltage controller.

Many different methods have been considered as the basis of analysing and designing a workable matrix converter. Because of the complexity of the necessary switching, the associated control logic is also complex and involves large and complicated algorithms. General requirements for generating PWM control signals for a matrix converter on-line in real time are:

- Computation of the switch duty cycles within one switching period.
- Accurate timing of the control pulses according to some predetermined pattern.
- Synchronisation of the computation process with the input duty cycle.

- Versatile hardware configuration of the PWM control system, which allows any control algorithm to be implemented by means of the software.

Two principal methods have been reported for the control of a matrix converter, being the Venturini Control Method and the Space Vector Modulation Control Method.[2] Matrix converters have not yet (2010) made any impact into the commercial converter market but the device remains an intriguing possibility for wind-energy applications.

7.7 Worked Numerical Examples[1]

7.7.1 *Three-phase bridge rectifier*

Example 7.1. A three-phase, full-wave, and controlled bridge rectifier contains six ideal thyristor switches and is fed from an ideal three-phase voltage source of 240-V and 50-Hz rating. The load is connected to the output terminals via a large smoothing inductor. Calculate the average load voltage at values of firing angle $\alpha_R = 30°, 60°,$ and $90°$.

-.-.-.-.-.-.-.-.-.-.-.-.-.-.-.-.-.-.-

As is customary the voltage rating, in this case 240 V, represents the rms value of the line-to-line voltage. The peak phase voltage E_m is given by:

$$E_m = \sqrt{2} \cdot \frac{240}{\sqrt{3}} = 196 \text{ V}.$$

The average voltage E_{av}, from Eq. (7.8), is therefore:

$$E_{av} = \frac{3\sqrt{3}}{\pi} E_m \cos \alpha_R$$

$$= \frac{3\sqrt{3}}{\pi} \cdot \sqrt{2} \frac{240}{\sqrt{3}} \cos \alpha_R$$

$$= 324 \cos \alpha_R.$$

(i) at $\alpha_R = 30°$, $E_{av} = 280.6$ V.
(ii) at $\alpha_R = 60°$, $E_{av} = 162$ V.
(iii) at $\alpha_R = 90°$, $E_{av} = 0$.

For the rectifier–inverter system of Fig. 7.2 the value E_{av} is the DC link voltage V_{dc}. No information about the link current I_{dc} and the consequent power output $V_{dc}I_{dc}$ can be obtained without a knowledge of the inverter output current and power requirement.

Example 7.2. The three-phase bridge rectifier of Example 7.1 is required to deliver a power of 10 kW into the inverter load. What must be the necessary value of the link current I_{dc} when $\alpha_R = 30°, 60°,$ and $90°$?

-.-.-.-.-.-.-.-.-.-.-.-.-.-.-.-.-.-

In Eq. (7.12):

$$P_{link} = V_{dc}I_{dc}.$$

At $\alpha_R = 30°$, $V_{dc} = 280.6$ V, from Example 7.1:

$$I_{dc} = \frac{10{,}000}{280.6} = 35.64 \text{ A}.$$

At $\alpha_R = 60°$, $V_{dc} = 162$ V, from Example 7.1:

$$I_{dc} = \frac{10{,}000}{162} = 61.73 \text{ A}.$$

At $\alpha_R = 90°$, $V_{dc} = 0$.

The bridge cannot transmit power at all unless the firing angle is reduced in value.

7.7.2 Three-phase bridge inverter feeding on infinite bus

Example 7.3. Power is transferred from a 300-V battery to a three-phase, 230-V, and 50-Hz AC bus via a controlled SCR inverter. The inverter switches may be considered lossless. A large filter inductor with resistance 10 Ω is included on the DC side. Calculate the power transferred if (a) $\alpha_I = 90°$, (b) $\alpha_I = 120°$, and (c) $\alpha_I = 150°$

-.-.-.-.-.-.-.-.-.-.-.-.-.-.-.-.-.-

The circuit is represented in Fig. 7.7 with the AC side current presumed to have the waveform shown in Fig. 7.8(a). The average voltage of the inverter is given by expression (7.13):

$$E_{av} = \frac{3\sqrt{3}}{\pi} E_{mb} \cos \alpha_I.$$

Since the line-to-line value of the bus voltage is 230 V, the peak phase voltage E_{mb} is:

$$E_{mb} = \sqrt{2}\frac{230}{\sqrt{3}} = 187.8 \text{ V}.$$

Therefore, the average voltage for the three cases is:

$$E_{av} = \frac{3\sqrt{3}}{\pi} \times 187.8 \times \cos \alpha_I$$
$$= 310.6 \cos \alpha_I.$$

This becomes:

$$E_{av} = 0, \text{ at } \alpha_I = 90°$$
$$= -155.3 \text{ V at } \alpha_I = 120°$$
$$= -269 \text{ V at } \alpha_I = 150°.$$

The negative sign indicates that the average value E_{av} of the inverter voltage opposes the current flow that is created by the application of the driving direct voltage V_{dc}.

In terms of the circuit of Fig. 7.7:

$$I_{dc} = \frac{V_{dc} + E_{av}}{R_f} = \frac{|V_{dc}| - |E_{av}|}{R_f}.$$

At $\alpha_I = 90°$:

$$I_{dc} = \frac{300}{10} = 30 \text{ A}.$$

At $\alpha_I = 120°$:

$$I_{dc} = \frac{300 - 155.3}{10} = 14.47 \text{ A}.$$

At $\alpha_I = 150°$:
$$I_{dc} = \frac{300 - 269}{10} = 3.1 \text{ A}.$$

The power transferred through the inverter into the AC system is the battery input power minus the power loss in the filter resistor R_f.
$$P = V_{dc}I_{dc} - I_{dc}^2 R_f.$$

At $\alpha_I = 90°$:
$$P = 300 \times 30 - (30)^2 10$$
$$= 9000 - 9000 = 0.$$

At $\alpha_I = 120°$:
$$P = 300 \times 14.47 - (14.47)^2 10$$
$$= 4341 - 2093.8 = 2247.2 \text{ W}.$$

At $\alpha_I = 150°$:
$$P = 300 \times 3.1 - (3.1)^2 10$$
$$= 930 - 96.1 = 833.9 \text{ W}.$$

The peak height of the AC side current pulses is also the battery current I_{dc}.

The AC side current pulses have a fundamental harmonic component with an rms value I_{a_1} given by Eq. (7.18):
$$I_{a_1} = 0.78 I_{dc}$$

when $\alpha_I = 90°$, $I_{a_1} = 0.78 \times 30 = 23.4 \text{ A}$
when $\alpha_I = 120°$, $I_{a_1} = 0.78 \times 14.47 = 11.3 \text{ A}$
when $\alpha_I = 150°$, $I_{a_1} = 0.78 \times 3.1 = 2.42 \text{ A}$.

The power on the AC side is given by the general form of Eq. (7.17):
$$P = 3 E_b I_{a_1} \cos \alpha_I$$
$$= 3 \times \frac{230}{\sqrt{3}} I_{a_1} \cos \alpha_I$$
$$= 398.4 \, I_{a_1} \cos \alpha_I.$$

At $\alpha_I = 90°$:
$$P = 398.4 \times 23.4 \times \cos 90° = 0.$$

At $\alpha_I = 120°$:
$$P = 398.4 \times 11.3 \times \cos 120° = -2251 \text{ W}.$$
At $\alpha_I = 150°$:
$$P = 398.4 \times 2.42 \times \cos 150° = -834.9 \text{ W}.$$
These power values are seen to agree closely with the corresponding values calculated on the DC side, described above.

Example 7.4. For the battery driven inverter of the previous example, Example 7.3, calculate the operating power factor at the firing angles (a) $\alpha_I = 90°$, (b) $\alpha_I = 120°$, and (c) $\alpha_I = 150°$.

-.-.-.-.-.-.-.-.-.-.-.-.-.-.-.-.-.-.-

For a system with sinusoidal voltages, the power factor can be expressed in terms of a displacement factor and a distortion factor, as described in Sec. 7.4.5.

The displacement factor, $\cos \Psi_1$, is equal to $\cos \alpha_I$ from Eq. (7.16). In this case:

At $\alpha_I = 90°$; displacement factor $= |\cos 90°| = 0$.
At $\alpha_I = 120°$; displacement factor $= |\cos 120°| = 0.5$.
At $\alpha_I = 150°$; displacement factor $= |\cos 150°| = 0.866$.

The distortion factor of the waveform in Fig. 7.16(a) was shown in Eq. (7.31) to have the value $3/\pi = 0.95$.

But the PF is the product of the displacement factor and the distortion factor, as defined in Eq. (7.28).

At $\alpha_I = 90°$:
$$\text{PF} = \frac{3}{\pi} \times 0 = 0.$$
At $\alpha_I = 120°$:
$$\text{PF} = \frac{3}{\pi} \times 0.5 = 0.477.$$
At $\alpha_I = 150°$:
$$\text{PF} = \frac{3}{\pi} \times 0.866 = 0.827.$$

As an alternative to the use of the displacement factor and distortion factor, the PF may be obtained directly from the defining Eq. (7.27).

The rms value of the AC side current pulses is given by Eq. (7.26):
$$I_a = 0.82\, I_{dc}.$$
At $\alpha_I = 90°$; $I_a = 0.82 \times 30 = 24.6\,A.$
At $\alpha_I = 120°$; $I_a = 0.82 \times 14.47 = 11.87\,A.$
At $\alpha_I = 150°$; $I_a = 0.82 \times 3.1 = 2.54\,A.$

The value of the per-phase rms voltage is constant.
$$E = \frac{230}{\sqrt{3}} = 132.8\,V.$$
Therefore, from Eq. (7.27):
$$PF = \frac{P}{3 E_a I_a}.$$

At $\alpha_I = 90°$:
$$PF = \frac{0}{3 \times 132.8 \times 24.6} = 0.$$

At $\alpha_I = 120°$:
$$PF = \frac{2251}{3 \times 132.8 \times 11.87} = 0.48.$$

At $\alpha_I = 150°$:
$$PF = \frac{834.9}{3 \times 132.8 \times 2.54} = 0.83.$$

The values of power factor calculated directly from the definition Eq. (7.27) are seen to agree closely with the values obtained from the displacement factor times distortion factor method.

Example 7.5. Calculate the voltage and current ratings of the thyristor switches in the inverter operation described in Examples 7.3 and 7.4.

-.-.-.-.-.-.-.-.-.-.-.-.-.-.-.-.-.-

All of the thyristor switches undergo equal duty. The operation of the inverter (Fig. 7.7), requires that the peak value of the line-to-line voltage falls sequentially on each switch. Using the terminology $V_T(\max)$, it is found that:
$$V_T(\max) = \sqrt{2} \times 230 = 325.3\,V.$$
Each switch of the inverter carries a current in which the waveform may be represented by the positive part of the current pulses of Fig. 7.8(a).

Considering the first half-cycle, utilising the defining equation for the rms value (7.25), it is seen that:

$$I_T = \sqrt{\frac{1}{2\pi} \int_{\frac{\pi}{6}}^{\frac{5\pi}{6}} I_{dc}^2 \, d\omega t} = \sqrt{\frac{I_{dc}^2}{3}}.$$

Therefore, in general:

$$I_T = \frac{I_{dc}}{\sqrt{3}}.$$

The maximum value of I_{dc} occurs when $\alpha_I = 90°$ so that $I_T = 30/\sqrt{3} = 17.32$ A.

If the general expression for I_T is combined with Eq. (7.26), it is confirmed because it is found that $I_T = I_a/\sqrt{2}$. In other words, the rms value of the line (phase) current is given by $I_a = \sqrt{2} I_T$. The choice of switch rating, to allow factors of safety, might be 400 V and 25 A.

7.8 Commonly Used Forms of Power Electronic Drive in Wind-Energy Systems

Most of the European wind-energy turbines now (2010) installed are based on one of three main wind turbine systems:

1. fixed-speed and directly coupled cage induction generator,
2. variable-speed, doubly fed, and wound rotor induction generator, and
3. variable-speed and direct-drive system with a synchronous generator.

7.8.1 *Fixed-speed and directly coupled cage induction generator*

In this scheme, the turbine rotor is coupled to the induction machine cage rotor, via a gearbox (Fig. 7.17). The induction generator stator winding is directly connected to the electrical grid. An induction generator slip varies with generated power so that the speed is not strictly constant but undergoes variation of about 1–2%. Because of the leading vars delivered by the generator, interpreted in power system terms as a sink of lagging vars, the generator is almost always PF compensated by terminal capacitors.

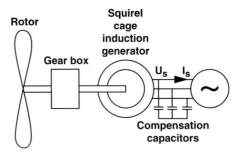

Fig. 7.17 Singly-fed induction generator.

A sophisticated variation of this system uses fast power electronics to create instantaneous variations of the rotor resistance. The commercial system uses the trade name OptiSlip and has been installed in turbines in the range 600 kW–2.75 MW. The use of the variable rotor (secondary) resistance permits transient speed increases up to 10% of the nominal value.

The fixed-speed wind-turbine systems are relatively simple in design and are slightly cheaper than the variable speed systems, as illustrated in Table 2.4 of Chap. 2. Because the rotor speed is fixed, while the wind speed varies, the fluctuations in wind speed create torque fluctuations in the power train. The resulting power fluctuation can cause variation of grid voltage resulting in lamp "flicker".

7.8.2 *Variable-speed and doubly fed induction generator*

An induction machine in which both the stator (primary) and the rotor (secondary) have three-phase windings can be operated as a doubly fed generator (DFIG). The wind-turbine rotor drives the induction machine rotor via a gearbox (Fig. 7.18). The induction machine stator is connected directly to the electrical grid. Decoupling is achieved between the fixed frequency of the grid and the variable (slip) frequency of the induction generator rotor by means of a power electronic switching converter, making variable speed of generator rotation possible. It is common to use an inverter switching logic that produces output voltage waveforms known as pulse-width modulation (PWM).

The doubly fed connection has been extensively used as an energy-saving device in induction motor drives, where it is usually described as

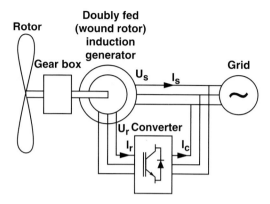

Fig. 7.18 Doubly-fed induction generator (DFIG).

"slip-energy recovery".[1] The power electronic converter in the rotor circuit usually needs to be rated lower than the generator rating, which is a cost saving on the expensive solid-state switches used. Depending on the speed range used, the power converter might be rated at about only one-third the rating of the generator. For example, if the drive speed range is 0.75–1.25 times synchronous speed, then the converter needs to be rated at only 0.25 times the machine rating. The system is very versatile and permits independent control of both the active and the reactive powers.

Because of the effective decoupling, the rotor can act as an energy storage system, absorbing the torque pulsations created by wind gusts. There is a requirement for periodic slip-ring maintenance in the rotor circuit.

7.8.3 Variable-speed and direct drive synchronous generator

In the scheme shown in Fig. 7.19, the synchronous generator rotor is mechanically coupled directly to the wind turbine. The direct drive design requires a relatively heavy generator and a larger rated power electronic converter, through which 100% of the generated power must pass, compared with the doubly fed induction generator described in the previous sub-section.

In an induction generator, all of the excitation is provided from the stator winding. The necessary level of air-gap flux density can be provided,

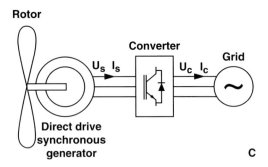

Fig. 7.19 Variable-speed, direct drive system using a synchronous generator.

from the stator, by the use of a relatively small air gap. By comparison, synchronous machines have their excitation systems on the rotor and can operate with larger air gaps.

7.9 Problems and Review Questions

7.9.1 *Three-phase controlled bridge rectifier, with ideal supply, feeding a highly inductive load*[1]

7.1 A three-phase, full-wave, controlled bridge rectifier contains six ideal thyristor switches and is fed from an ideal three-phase supply of balanced sinusoidal voltages. The load consists of a three-phase inverter fed via a DC link containing a large filter inductor. Show that, for all values of thyristor firing angle α_R, the load voltage is given by Eq. (7.8).

7.2 A three-phase, full-wave, controlled bridge rectifier has ideal switches and is supplied from an ideal three-phase voltage supply rated at 415 V and 50 Hz. The load includes a series connected filter inductor. Calculate the average load voltage at the firing angles (a) $\alpha_R = 0$, (b) $\alpha_R = 30°$, (c) $\alpha_R = 60°$, and (d) $\alpha_R = 90°$.

7.3 Show that for the inductively loaded bridge of Problem 7.1, the distortion factor of the supply current is independent of the thyristor firing angle.

7.4 Show that the waveform of the supply current to a three-phase, full-wave, controlled bridge rectifier with highly inductive load is given by:

$$i(\omega t) = \frac{2\sqrt{3}I_{dc}}{\pi}\left[\sin(\omega t - \alpha) - \frac{1}{5}\sin 5(\omega t - \alpha)\right.$$
$$\left. - \frac{1}{7}\sin 7(\omega t - \alpha) \cdots\right],$$

where I_{dc} is the average load current.

7.5 For the inductively loaded bridge rectifier of Problem 7.1, show that the rms supply current is given by:

$$I = \frac{3\sqrt{2}}{\pi R}E_m \cos\alpha_R.$$

7.6 For the inductively load bridge of Problem 7.1, calculate the rms current and peak reverse voltage ratings required of the bridge switches.

7.9.2 Three-phase, full-wave, and controlled bridge inverter feeding an infinite bus[1]

7.7 A naturally commutated, three-phase inverter contains six ideal switches and transfers energy into a 440-V, 50-Hz, three-phase supply from an 800-V DC battery. The battery and the inverter are linked by a smoothing inductor with a resistance of 12.4 Ω. Calculate the power transferred at $\alpha_I = 90°$, $120°$, $150°$, and $170°$.

7.8 For the inverter application of Problem 7.7, calculate the voltage and rms current ratings required of the switches.

7.9 A large solar energy installation utilises a naturally commutated and three-phase SCR inverter to transfer energy into a power system via a 660-V transformer. The collected solar energy is initially stored in an 800-V battery that is linked to the inverter through a large filter choke of resistance 14.2 Ω. What is the maximum usable value of the SCR firing angle? Calculate the power transferred at the firing angle of 165°. What is the necessary transformer rating?

7.10 Calculate the necessary SCR voltage and rms current ratings for the inverter application of Problem 7.9.

7.11 Use the inverter characteristics of Fig. 7.18 to deduce the form of the corresponding V_{dc}–I_{dc} characteristics with $\cos\alpha_I$ as the parameter.

If the maximum DC-side voltage is 100 V, what is the firing angle required to give a DC of 10 A if $R = 1\,\Omega$?

7.12 Sketch the main circuit of a naturally commutated, three-phase, and controlled bridge inverter. If the AC-side rms line voltage E is fixed, sketch the variation of inverter power transfer with SCR firing angle α and DC-side voltage V_{dc}. If $\alpha_I = 120°$, what is the minimum value of ratio V_{dc}/E that will permit inversion? Sketch the waveform of the current passing between the inverter and the AC supply and give a phasor diagram interpretation to explain the inverter operation. Why is it necessary (usually) to connect capacitance across the output terminals of a naturally commutated inverter of high kVA rating?

References

1. Shepherd, W., L. N. Hulley, D. T. W. Liang, *Power Electronics and Motor Control*, Second Edition, Cambridge University Press, Cambridge, England, 1995.
2. Shepherd, W., L. Zhang, *Power Converter Circuits*, Marcel Dekker, Inc., New York, the USA, 2004.
3. Adamson, C., N. G. Hingorani, *High Voltage Direct Current Power Transmission*, Garraway Ltd., London, England, 1960
4. Electrical Issues Associated with Variable Speed Operation of Wind Turbines, ETSU W/33/00417/REP Contract Report from Garrad Hassan and Partners Ltd., UK, 1996.

CHAPTER 8

Integrating Wind Power Generation into an Electrical Power System

The wind blows intermittently. Irregularity of the wind is reflected into an irregular supply of energy from wind generators. This means that no electricity supply can rely wholly on wind generation but requires backup from a regular and reliable source.

In addition to the time-diffuse nature of the wind, the individual wind turbines and wind farms are often distributed over wide geographical areas. The sources of wind-powered generation are therefore scattered over wide areas but need to plug into the electrical power network close to their points of generation.

The power ratings of wind power installations for electricity generation vary from individual turbines at (say) 25 kW to large wind farms of up to 30 MW.[1] Transmission voltages on electrical power networks may be several hundred kilovolts in rating. This is much too high for customer distribution so the transmission voltages are transformed down to a few kilovolts. Wind turbines and wind farms are not usually connected to the transmission grid but to the lower voltage distribution network and are said to be "embedded" into the distribution system.[1]

8.1 Electricity Distribution Systems

Most countries have electricity distribution systems that are similar in basic form. The essential components for a system in the United Kingdom (UK) with typical UK voltage values are shown in Fig. 8.1.

The generator voltage is transformed up to 400 kV, 275 kV or 132 kV for high-voltage transmission. A high transmission voltage permits a lower transmission current and therefore reduced transmission losses. At the receiving end of the high voltage line, the voltage level is transformed down to 33 kV and the 11 kV for the distribution system, which is much smaller in distance than the transmission system. An earthing transformer with an interconnected star winding is usually connected at the distribution voltage level to ground the neutral points directly, or through a low-value resistor. A typical resistance value of 6.35 Ω permits 1,000 A to flow into a phase-ground fault.[2] Industrial loads are often supplied to factory substations at 11 kV or at 6.6 kV. For domestic supply in the UK the voltage level is further reduced to 400 V and three phase, which gives nominally 230 V and one phase. An incoming wind generator array is normally connected into the system at the 11-kV level, or sometimes at 33 kV. All of the voltage levels are chosen for economic reasons, bearing in mind the operating levels needed for the system protection devices and the need for safe operation at all times.

The frequency of the system is controlled at the main generation points. This is nominally 50 Hz in the UK and 60 Hz in North America. All countries

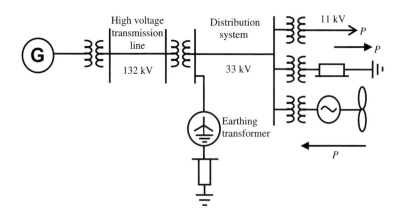

Fig. 8.1 Basic form of an electricity distribution system.

have strict regulation of the system frequency. In the UK, the frequency variation permitted by law is in the range of ±0.5%. In addition, in the UK, there are strict legal limits permitted of system voltage variation. In 11-kV circuits, the permitted variation is ±1% or ±2%. Higher voltage circuits at 33 kV or 132 kV are permitted variations of ±6% because automatic on-load tap changers can compensate for variations in the network voltage.[2]

8.2 Issues for Consideration Concerning the Integration of Wind-Energy Generation into an Electric Power System

Two basic questions immediately arise when considering the connection of a wind generator or a wind farm into an established and working power network.

1. What is the best point of coupling into the existing grid?
2. How much wind-generation capacity can be connected into the network at any given point?

Neither of these questions can be answered by rigorous mathematical analysis or by the use of standard formulae. Both issues depend on the electrical nature of the network and especially on the required load flow of both real and reactive powers. In addition, both issues depend on wind supply, which can be defined only in terms of statistical probability.

The connection of a significant level of wind generation into a system will have effects throughout all parts of the network. The following list of issues is quoted from Ref. [1].

From the perspective of the main generation, corresponding to generator G in Fig. 8.1, the main issues to be addressed are:

- energy credit,
- capacity credit, and
- control issues.

From the perspective of the transmission system, there arise issues of:

- operating costs and
- equipment requirements.

From the perspective of the network distribution system, there arise issues of:

- operating costs,
- equipment requirements,
- power quality, and
- protection and safety.

An overall assessment as to whether the connection of wind energy represents an advantage involves consideration of both technical and economical factors. Further consideration of some of the above issues is given in following sections of this chapter.

8.2.1 *Energy credit*

The major benefit from the use of wind energy is the saving on fossil or nuclear fuels that would otherwise be required. For large power systems with only a small proportion of wind turbine contribution (usually called a "small penetration"), the energy output of a wind farm is approximately equal to the energy not required from the conventional plant. Each kWh of energy generated by the wind turbines results in one kWh of output saved from the conventional generation. Wind generation is expected to increase in penetration in the future and when this exceeds 5%, the effect of wind energy on the variation of the load seen by conventional plant will become important and fuel savings will be reduced.[1] At present (2009), few grid systems have the penetration of wind energy above 5%, the exceptions being Denmark (18%), Spain and Portugal (9%+), Germany (6%+), and the Republic of Ireland (6%+). The Danish grid is heavily interconnected to the European electrical grid and it has solved grid management problems by exporting almost one-half of its wind power generation to Norway. A study commissioned by the US state of Minnesota has considered penetration of up to 25% and concluded that integration issues would be manageable and have incremental costs of less than one-half cent ($0.0045) per kWh.[3,4] The growth of wind energy contribution and the increasing energy credit that accrues are illustrated by the Table of Installed Wind Power Capacity (MW) shown as Table 8.1.[4]

Table 8.1 Installed wind power capacity (MW)[4].

Rank	Nation	2002	2003	2004	2005[2]	2006[2]	2007[2]	2008[2]	2009[3]	1 Yr % growth	5 Yr avg % growth
-	World	31,180	39,295	47,693	59,024.1	74,150.8	93,926.8	121,187.9	157,899	30.3	25.3
-	European Union				40,722	48,122	56,614	65,255	74,767	12.1	
1	United States	4,685	6,370	6,725	9,149	11,603	16,818.8	25,170.0	35,159	39.7	31.6
2	Germany	12,001	14,609.1	16,628.8	18,427.5	20,622	22,247.4	23,902.8	25,777	7.8	10.3
3	China	468	567	764	1,266	2,599	5,912	12,210.0	25,104	105.6	84.8
4	Spain	4,830	6,202	8,263	10,027.9	11,630	15,145.1	16,740.3	19,149	14.4	22.0
5	India	1,702	2,110	3,000	4,430	6,270	7,850	9,587.0	10,925	14.0	35.4
6	Italy	785	904	1,265	1,718.3	2,123.4	2,726.1	3,736.0	4,850	29.8	32.8
7	France	148	248	386	757.2	1,567	2,455	3,404.0	4,410	29.6	68.9
8	United Kingdom	552	648	888	1,353	1,962.9	2,389	3,287.9	4,070	23.8	38.4
9	Portugal	194	299	522	1,022	1,716	2,130	2,862.0	3,535	23.5	57.1
10	Denmark	2,880	3,110	3,124	3,128	3,136	3,125	3,160.0	3,465	9.7	0.3
11	Canada	236	322	444	683	1,460	1,846	2,369.0	3,319	40.1	49.1
12	Netherlands	682	908	1,078	1,224	1,559	1,747	2,225.0	2,229	0.0	19.6
13	Japan	334	506	896.2	1,040	1,309	1,528	1,880.0	2,056	9.4	30.0
14	Australia	103	197.2	379	579	817.3	817.3	1,494.0	1,712	14.6	49.9
15	Sweden	345	404	452	509.1	571.2	831.0	1066.9	1,560	28.4	21.4
16	Ireland	137	186	338.9	495.2	746	805	1,244.7	1,260	54.6	46.3
17	Greece	276	365	473	573.3	757.6	873.3	989.7	1,087	13.3	22.1
18	Austria	139.3	415	606	819	964.5	981.5	994.9	995	1.4	19.1
19	Turkey	19.4	20.6	20.6	20.1	64.6	206.8	333.4	801	61.2	74.5
20	Poland	28.2	58.2	58.2	73	153	276	472.0	725	71.0	52.0
21	Brazil	22	23.8	23.8	28.6	236.9	247.1	338.5	606	37.0	70.1
22	Belgium	44	68	95	167.4	194.3	286.9	383.6	563	33.7	41.3
23	New Zealand	35	36.3	168.1	168.2	171	321.8	325.3	497	1.1	55.1
24	Taiwan	0	0	13	103.7	187.7	279.9	358.2	436	28.0	129.0[4]
25	Norway	97.3	100	270	268	325	333.0	428.0	431	28.5	33.7
26	Egypt	69	180	145	145	230	310	390.0	430	25.8	16.7
27	Mexico	0	0	2.2	2.2	84	85.0	85.0	402	0	
28	South Korea	16	18.7	22.5	119.1	176.3	192.1	278.0	348	44.7	71.6
29	Morocco	53.9	53.9	53.9	64	64	125.2	125.2	253	0	18.4
30	Hungary	1.2	2	3.5	17.5	60.9	65.0	127.0	201	95.4	129.4
31	Czech Republic	3	10	16.5	29.5	56.5	116	150.0	192	29.3	71.9
32	Bulgaria	0	10	10	14	36	56.9	157.5	177	176.8	73.6

The wind-generation capacity of the world more than quadrupled between 2000 and 2006. Now, 81% of the wind power installations are in the USA and Europe, but the share of the top five countries in terms of new installations fell from 71% in 2004 to 62% in 2006.

By 2010, the World Wind Energy Association expects 160 GW of capacity to be installed worldwide, up from 73.9 GW at the end of 2006, implying an anticipated net growth rate of more than 21% per year.

Denmark generates nearly one-fifth of its electricity with wind turbines — the highest percentage of any country — and is fifth in the world in total wind-power generation. Denmark is prominent in the manufacturing and use of wind turbines, with a commitment made in the 1970s to produce eventually one-half of the country's power by wind power.

Germany was the leading producer of wind power, with 28% of the total world capacity in 2006 and a total output of 38.5 TWh in 2007 (6.3% of German electricity); the official target is for renewable energy to meet 12.5% of German electricity needs by 2010 — this target may be reached ahead of schedule. Germany has 18,600 wind turbines, mostly in the north of the country — including three of the biggest in the world, constructed by the companies Enercon (6 MW), Multibrid (5 MW), and Repower (5 MW). Germany's Schleswig-Holstein province generates 36% of its power with wind turbines.

In recent years, the USA has added more wind energy to its grid than any other country; US wind-power capacity grew by 45% to 16.8 GW in 2007. Texas has become the largest wind-energy-producing state, surpassing California. In 2007, the state expected to add 2 GW to its existing capacity of approximately 4.5 GW. Iowa and Minnesota were expected to produce each 1 GW by late 2007. Wind-power generation in the USA was up 31.8% in February 2007 from February 2006. The average output of 1 MW of wind power is equivalent to the average electricity consumption of about 250 American households. According to the American Wind Energy Association, wind would generate enough electricity in 2008 to power just over 1% (4.5 million households) of total electricity in the USA, up from less than 0.1% in 1999. US Department of Energy studies have concluded that wind harvested in just three of the fifty US states could provide enough electricity to power the entire nation, and that offshore wind farms could do the same job.

India ranks fifth in the world with a wind-power capacity of 10925 MW in 2009, or 3% of all electricity produced in India. The World Wind Energy Conference in New Delhi in November 2006 has given additional impetus to the Indian wind industry. The wind farm near Muppandal, Tamil Nadu, India, provides an impoverished village with energy. India-based Suzlon Energy is one of the world's largest wind-turbine manufacturers.

In December 2003, General Electric installed the world's largest offshore wind turbines in Ireland, and plans are being made for more such installations on the west coast, including the possible use of floating turbines.

In 2005, China announced that it would build a 1,000 MW wind farm in Hebei for completion in 2020. China reportedly has set a generating target of 20,000 MW by 2020 from renewable energy sources — it says indigenous wind power could generate up to 253,000 MW. Following the World Wind Energy Conference in November 2004, organised by the Chinese and the World Wind Energy Association, a Chinese renewable energy law was adopted. In late 2005, the Chinese government increased the official wind energy target for the year 2020 from 20 to 30 GW.

Canada experienced rapid growth of wind capacity between 2000 and 2006, with total installed capacity increasing from 137 to 1,451 MW, and showing an annual growth rate of 38%. Particularly rapid growth was seen from 2006, with the total capacity more than doubling to the end of 2009. This growth was fed by measures including installation targets, economical incentives, and political support. For example, the Ontario government announced that it will introduce a feed-in-tariff for wind power, referred to as "Standard Offer Contracts", which may boost the wind industry across the province. In Quebec, the provincially owned electric utility plans to purchase an additional 2,000 MW by 2013.

8.2.2 Capacity credit

The installation of a wind-energy component into the power-generation system usually means that there is then reduced need for conventional thermal generation. The capacity credit is the fraction of the installed wind capacity by which the conventional power-generation capacity can be reduced without affecting the reliability of the system.[5,6]

There are two different principal methods of calculating the capacity credit:

1. The trade-off between conventional and wind plant is often based on a reliability measure known as loss-of-load expectation or LOLE. This is an indication of the statistically expected number of times within a given time period that the system could not provide for customer load. If the wind-generating capacity can be substituted for the conventional capacity, without the reduction of system reliability, then the wind plant capacity credit can be calculated.
2. An alternative to the LOLE method is to examine the overall economics of wind-power plants by using production-cost models to estimate the benefits. The focus of the latter method is the wind-energy displacement value, over several years of operation. The contribution made by wind plant at times of peak demand is assessed and the average value taken as the capacity credit.[1]

The accuracy of a capacity credit rating depends on the accuracy of wind forecasting. But, although the accurate wind forecasts help with individual unit commitment, they may also significantly reduce energy imbalance charges.

"Both of the methods yield very similar results. For small penetrations of wind power the capacity credit is usually close to the average output of the wind plant, i.e., 1,000 MW of wind plant generally has a capacity credit of about 300 MW".[1]

8.2.3 Control and reliability[7]

An integrated power system is large and very complex, containing thousands of individual components and machines. The control issue is to adjust continually the supply of electrical energy to satisfy the continually changing demand of the system users. High reliability has to be maintained at the lowest cost. The physical limitations of individual network elements such as generators, transmission lines, and substations must be observed at all times.

The system operation involves:

- Keeping the voltage at each node within prescribed limits.

- Regulation of the system frequency to keep all generating units, including the wind generation, in synchronism.
- Ensure that the system can withstand and recover from unplanned failures or the loss of major elements.

The following items describe the range of technical aspects to be considered to maintain reliable operation of the system.

- Voltage regulation and reactive power dispatch. This involves the deployment of machines or/and devices capable of generating reactive power.
- Frequency regulation by adjusting certain generating units in response to fast fluctuations in the total system load.
- Adjusting the generation and load dispatching to follow the daily load demand patterns.
- Maintaining an adequate level of spinning reserve (usually on-line and synchronised to the system) that can quickly respond to the loss of a major transmission element, such as a wind generator.
- Managing an additional back-up supply of generating capacity that can be brought on-line relatively quickly in the event of a major unplanned deficiency of wind (or other) supply.

The basis of reliable system operation is accurate load forecasting, even though this can never be 100% accurate. Sophisticated load-forecasting techniques and practical experience both contribute to the reduction of uncertainty. Although the performance of an individual wind turbine is subjected to much uncertainty due to wind variability, the aggregate performance of a large number of turbines is of primary interest with regard to the impact on the transmission grid and system operation. The existence of several wind farms in different locations contributes a smoothing effect to the wind generation level available. Moreover, the greater the distances between different connected wind farms, the greater is the smoothing.

In the UK, the National Grid Transco has summarised the key issues related to smoothing as follows:

> However, based on recent analysis of the incidence and variation of wind speed we have found that the expected intermittency of wind does not pose such a major problem for stability and we are confident that this can be adequately managed. ... It is a property of the interconnected transmission system that individual and local independent fluctuations in output are diversified and averaged out across the system.[8]

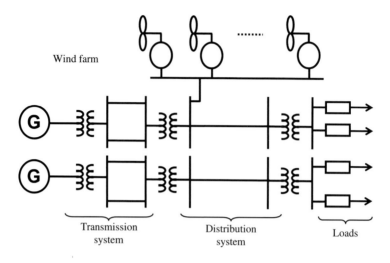

Fig. 8.2   Electrical power system incorporating a wind farm.

8.3 The Effect of Integrated Wind Generation on Steady-State System Voltages

As power flows through an electrical supply system, the voltages at various points (often called "nodes") vary. A conceptual line diagram, Fig. 8.2, incorporates a transmission system, a distribution system to which a wind farm is connected, and a series of loads. The system voltages at the various transformer terminals have to be maintained within design limits, sometimes with the assistance of automatic voltage regulation. Most power system impedances, due to transmission lines and transformers, are resistive–inductive in nature and can be represented by series resistance–inductance elements in an equivalent circuit. A radial circuit connecting a wind turbine to a fixed voltage point is shown in the equivalent circuit of Fig. 8.3.

In the wind-generator system of Fig. 8.3, the induction generator delivers real power P into the fixed voltage bus of the power system at a voltage V_b. The induction generator delivers leading reactive volt-amperes (VARs) Q into the system, at a power factor (usually) of about 0.9 leading, as described in detail in Sec. 6.4.2 of Chap. 6. A capacitor bank at the generator terminals absorbs or partly absorbs the VARs delivered by the induction generator.

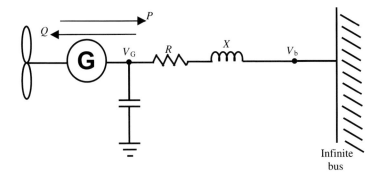

Fig. 8.3   Fixed-speed wind generator connected to a fixed voltage point.

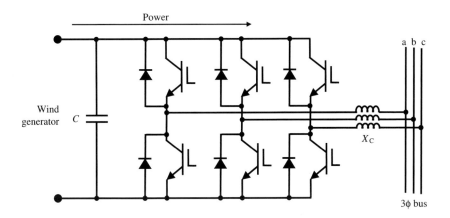

Fig. 8.4   Three-phase, PWM voltage source inverter using IGBT switches.[2]

To be consistent with the European definition of complex power Eq. (5.36) of Chap. 5, the flow of the reactive power Q in power systems is thought of as being lagging reactive power. Accordingly, the flow of leading VARs from an induction generator or an underexcited synchronous generator is interpreted, for load-flow calculations purposes, as being a load of lagging VARs.

The value of the variable generator voltage V_G can be determined in terms of impedance and power variables and the fixed system voltage V_b, from the relationship:[9]

$$V_G^4 + V_G^2[2(QX + PR) - V_b^2] + (QX + PR)^2 + (PX - QR)^2 = 0. \quad (8.1)$$

For lightly loaded distribution circuits, the voltage difference across the link can be approximated as:

$$V_G - V_b = \frac{PR - QX}{V_b}. \qquad (8.2)$$

An increase of real power transfer P through the link tends to increase the voltage drop (i.e., increase the wind-turbine voltage) whereas an increase of reactive power Q tends to reduce the voltage drop.

The flow of current through the link circuit resistance R in Fig. 8.3 causes a power loss in the circuit that can be approximated by:

$$P_{loss} = \frac{(P^2 + Q^2) R}{V_b^2}. \qquad (8.3)$$

Equation (8.3) does not have the dimension of power but is only a calculation device in which the components P, Q, R, and V_b are all per-phase quantities.

Most practical power system line diagrams are much more complicated than the circuit of Fig. 8.3 and can contain dozens, or even hundreds, of branches. These circuits cannot be solved using simple relationships such as those of Eqs. (8.2) and (8.3). Analysis of the load flow of P and Q involves many simultaneous equations that can be solved only by numerical iteration methods such as the Gauss-Siedel method. The interested reader is referred to the various excellent texts on power system analysis.

Example 8.1 (adapted from Ref. [1])

A wind turbine is rated at 750 kW. When delivering its rated load, it absorbs the equivalent of 240 kVARs. The turbine delivers power to a three-phase and 50-Hz bus rated at 415 V (line-to-line). The connecting link has the impedance dimensions $R = 0.007\ \Omega$/phase and $X = 0.0157\ \Omega$/phase, referred to 415-V operation. Calculate the generator terminal voltage and the power lost in the link.

—.—.—.—.—.—.—.—.—.—

The system shown in Fig. 8.3 applies directly here.

$$V_b = 415\ \text{V line-to-line}$$
$$= \frac{415}{\sqrt{3}} = 240\ \text{V/phase}.$$

In Eq. (8.2), the following parameter values apply.

$$P = \frac{750}{3} = 250 \, \text{kW/phase}$$
$$Q = \frac{240}{3} = 80 \, \text{kVAR/phase}$$
$$R = 0.007 \, \Omega/\text{phase}$$
$$X = 0.015 \, \Omega/\text{phase}.$$

In Eq. (8.2):

$$\begin{aligned}
V_G &= 240 + \frac{(250 \times 10^3 \times 0.007 - 80 \times 10^3 \times 0.015)}{240} \\
&= 240 + \frac{(1750 - 1200)}{240} \\
&= 240 + \frac{550}{240} \\
&= 240 + 2.29 = 242.3 \, \text{V/phase} \\
&= \sqrt{3} \times 242.3 = 420 \, \text{V line-to-line}.
\end{aligned}$$

The power loss per phase due to $I^2 R$ loss is given by Eq. (8.3).

$$\begin{aligned}
P_{\text{loss}} &= \frac{(250^2 + 80^2) \times 10^6 \times 0.007}{240^2} \\
&= 8400 \, \text{W/phase} \\
&= 3 \times 8400 = 25.2 \, \text{kW total},
\end{aligned}$$

which is 3.33% of the generator rating.

8.4 The Effect of Integrated Wind Generation on Dynamic and Transient System Voltages

All electric power systems, whatever the form of generation, are subject to sudden changes of state that result in voltage transients. The commonest form of change is due to switching events that might involve the switching in or out of major components such as a large load, transmission line, or generator. Such changes create immediate differences in the system impedances or even in the level of power and/or the reactive power demand. The switching event itself causes a voltage transient at the switch that might take the form of a sharp spike of voltage or an abrupt change of level that is accompanied by an oscillatory decay or rise of the voltage.

8.4.1 Lightning strikes

Electric power systems are subject to lightning strikes on the transmission and distribution equipment. These can cause serious damage, amounting to destruction, of equipment and can be hazardous to personnel. A lighting surge can have a very steep wave front with a rise time of about $1\,\mu$s, and is typically less than $10\,\mu$s. The resulting travelling wave of voltage passes through the system unless it is arrested or diverted to earth (ground) by protection equipment. Because a lightning voltage may be as high as several megavolts, the effect of a strike is likely to be very serious.

The protection of wind turbines and wind farms raises particular problems. The turbines are usually located at the tops of high towers and land-based machines are often on hilltops that contain high-resistivity ground. A wind-farm earthing system is required to work effectively both for power frequency 50/60-Hz currents and for lightning strikes that contain high-frequency components. The detail of the earthing practices used is given in various references but is outside the scope of the present text.[2,10,11]

8.4.2 Voltage flicker

For each rotation of a wind turbine, the propeller blades pass through the wind "shadow" of the tower, for both upstream and downstream designs. This causes a reduction of the wind force on the blades. In each rotation, there is a bending moment on the blade during the effective period of the shadow. At typical rotational speeds for high-power turbines, the blade passing frequency, which is the rotor speed times the number of blades, is usually of the order 1–2 Hz, or smaller. This action also causes a cyclic-flexing force at the root of the blades that could eventually result in fatigue failure of a propeller blade if the frequency and magnitude of the bending force was appropriate.

Cyclic variation of the force on the propeller blades represents a small variation on the torque supplied to the turbine and appears as a small ripple of the generator output voltage. If the generator was supplying an electrical load of incandescent lamps, the voltage ripple would be evident in flicker of the light from the lamps. Even at a generated frequency of 25 Hz, which was at one time used in some North American cities, this flicker is evident to the human eye and can be very irritating. The human eye appears to be most sensitive to variations of about 10 Hz.

The term "flicker" has been imported into wind engineering to describe the cyclic voltage perturbations created by wind tower shadow effects and other switching events. Flicker standards are often used to characterise transient voltage variations. Many countries have established standards for quantifying flicker and hold limits for allowable flicker and step changes in voltage.[12]

> The magnitude of the flicker due to wind turbulence depends on the slope of the real power v. reactive power characteristics of the generator, the slope of the power v. wind characteristics of the turbine, the wind speed and turbulence intensity. Flicker is, in general, much less of a problem for fixed-pitch, stall-regulated machines than for pitch-regulated machines.[12] Because embedded wind generators may start and stop frequently, the consequent flicker effects may become significant.

A much more analytical treatment of flicker is given in the book by Lubosny.[13]

8.4.3 Harmonics

When the frequencies of the voltages generated by a wind turbine-generator system are variable, as is common with variable-speed turbines, it is necessary to introduce a frequency converter between the generator and the system to which it is connected. In Fig. 6.22 and 6.23 of Chap. 6, for example, the two variable speed generator options are shown as part of systems that incorporate power electronic switching frequency converters.

The principal component of the frequency converter is the three-phase bridge voltage source inverter. A version of this using silicon-controlled rectifiers as electronic switches is shown in Fig. 7.7 and the action is described in Sec. 7.3 of Chap. 7. More modern systems use the insulated gate bipolar transistor (IGBT) as a switch and operate a switching strategy known as pulse-width modulation (PWM) to fabricate output voltage waveforms with little or no harmonic components lower that $n = 19$. At this harmonic order, the inductive components of the system impedance offer a much greater opposition to the flow of harmonic currents than is the case with the simpler form of inverter creating low-order harmonics due to step-wave voltages. It is desirable to minimise the flow of harmonic currents, as these increase the power loss in conductors and can cause the spurious tripping of protection relays.

The presence of harmonics in the system voltages and currents is manifested in non-sinusoidal waveforms. This is so important an issue that electricity authorities lay down strict guidelines, legally enforceable, on the levels of harmonics that are permissible at key points in the system.

Harmonic distortion is often assessed in terms of a property known as the total harmonic distortion (THD). Any non-sinusoidal repetitive waveform that is periodic with angular frequency ω can be represented as a summation of sinusoidal (or cosinusoidal) components:

$$\begin{aligned} v(\omega t) &= \frac{a_o}{2} + \sum_{n=1}^{\alpha} (a_n \cos n\omega t + b_n \sin n\omega t) \\ &= \frac{a_o}{2} + \sum_{n=1}^{\alpha} c_n \sin(n\omega t + \psi_n). \end{aligned} \qquad (8.4)$$

If the periodicity is 2π radians, then:

$$\begin{aligned} \frac{a_o}{2} &= \frac{1}{2\pi} \int_0^{2\pi} v(\omega t) d\omega t \\ &= \text{zero frequency component} \\ &= \text{DC component} \\ &= \text{time average value.} \end{aligned} \qquad (8.5)$$

$$a_n = \frac{1}{\pi} \int_0^{2\pi} v(\omega t) \cos n\omega t \cdot d\omega t. \qquad (8.6)$$

$$b_n = \frac{1}{\pi} \int_0^{2\pi} v(\omega t) \sin n\omega t \cdot d\omega t. \qquad (8.7)$$

It is found that:

$$\begin{aligned} c_n &= \text{peak value of the } n\text{th harmonic component} \\ &= \sqrt{a_n^2 + b_n^2}. \end{aligned} \qquad (8.8)$$

$$\begin{aligned} \psi_n &= \text{phase displacement of the } n\text{th harmonic component} \\ &= \tan^{-1}\left(\frac{a_n}{b_n}\right). \end{aligned} \qquad (8.9)$$

$$a_n = c_n \sin \psi_n. \qquad (8.10)$$

$$b_n = c_n \cos \psi_n. \qquad (8.11)$$

It is well to remember that although the Fourier series defined above is a brilliant mathematical theory and very useful in electric circuit analysis, it is only a theory. The Fourier harmonics are only mathematical constructs

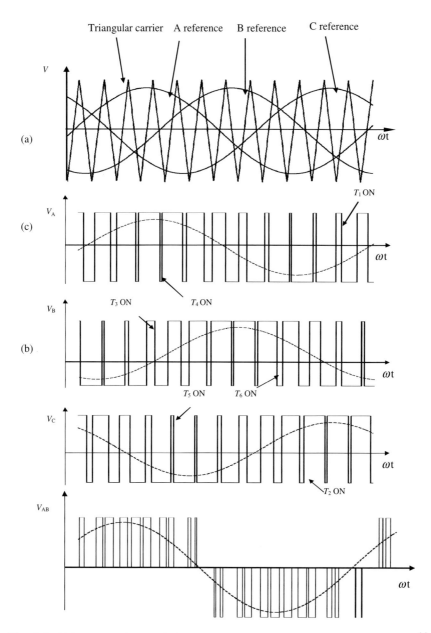

Fig. 8.5 Voltage waveforms of three-phase PWM inverter with natural sampling:[14] (a) reference signals for the IGBT gating; (b) phase voltages and fundamental (system frequency) components; and (c) line-to-line voltage waveform V_{AB} and its system frequency component.

and do not exist physically. You cannot see (say) the fundamental voltage components of Fig. 8.5 physically because they do not exist.

The THD is the ratio of the total energy in a waveform at all harmonic frequencies compared with the total energy at the system (fundamental) frequency. With AC voltages, there is no DC component so that $a_o = 0$ in Eq. (8.5). The fundamental component is represented by $n = 1$ and the sum of the harmonic frequencies is $\sum_{n=2}^{\alpha} v_n(\omega t)$. In mathematical terms:

$$\text{THD} = \left[\frac{\sum_{n=2}^{\alpha} \frac{1}{2\pi} \int_0^{2\pi} v_n^2(\omega t) d\omega t}{\frac{1}{2\pi} \int_0^{2\pi} v_1^2(\omega t) d\omega t} \right]^{1/2}. \quad (8.12)$$

For a pure sinusoid $v_1(\omega t) = v_n(\omega t)$ and the THD has the ideal value of zero. It is an aim of power system operation to make the THD as small as possible, preferably zero. Many electric power companies in Europe and in North America use the IEEE 519 standard to determine the maximum allowable THD for the voltage waveform at the point of common coupling (connection) or PCC.[15] This is quoted here in Table 8.2. Corresponding restrictions apply to the harmonic content of currents. In IEEE 519, the effective criterion is the ratio of maximum short-circuit fault current at the PCC.

Fixed-speed wind turbines using induction generators are sometimes connected into the system via thyristor soft-starting circuits. Each thyristor of the inverse-parallel connected pair (Fig. 8.6) is gated at the same angle delayed from its zero crossing to produce a symmetrical AC waveform. With resistive load, the single-phase load current would have the form shown in Fig. 8.7, for sinusoidal supply voltage. The fundamental component of the voltage waveform $v_1(\omega t)$ is found to be defined by the relationship, in terms

Table 8.2 Maximum allowable total harmonic distortion (THD) of voltage at the point of common coupling (PCC).

Voltage at point of coupling	Individual harmonic (%)	THD (%)
2.3–69 kV	3.0	5.0
69–138 kV	1.5	2.5
>138 kV	1.0	1.5

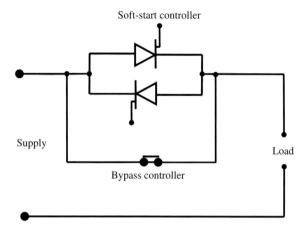

Fig. 8.6   Soft-start controller for limiting the load current.

of peak voltage V_{1m} and firing angle by:

$$v_1(\omega t) = V_{1\max} \sin(\omega t - \psi_1), \qquad (8.13)$$

where:

$$V_{1\max} = \frac{V_m}{2\pi}\sqrt{(\cos 2\alpha - 1)^2 + [\sin 2\alpha + 2(\pi - \alpha)]}. \qquad (8.14)$$

$$\psi_1 = \tan^{-1}\left[\frac{\cos 2\alpha - 1}{\sin 2\alpha + 2(\pi - \alpha)}\right]. \qquad (8.15)$$

When $\alpha = 90°$, for example:

$$v_1(\omega t) = 0.59\, V_m (\sin \omega t - 32.5°). \qquad (8.16)$$

The rms value of the load voltage smoothly reduces from its peak source value (V_m in Fig. 8.7) to zero as $\alpha \to \pi$. By phasing back the firing angle, the voltage applied to the load can be reduced to a low value and gradually increased as required. The waveform of Fig. 8.7 is characterised by low-order harmonics, the largest being the third harmonic of the source frequency. This form of single-phase circuit is extensively used for load voltage control in lighting dimmer circuits. A three-phase version is often used to connect the induction generators used with fixed-speed wind turbines and provide "soft" (i.e., reduced voltage) starting.

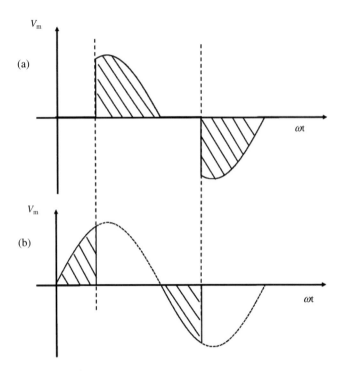

Fig. 8.7 Voltage waveforms of the two-thyristor soft-starter (Fig. 8.6) with resistive load and sinusoidal supply. Firing angle $= \alpha = 60°$: (a) load voltage and current and (b) voltage across the thyristor (SCR) switch.

8.4.4 Self-excitation of induction generators

With fixed-speed induction generators, the magnetising current drawn from the connected system, via the machine stator windings, is usually compensated by parallel-connected capacitors. A typical connection is shown in Fig. 6.21 of Chap. 6. The corresponding per-phase equivalent circuit is given in Fig. 6.17.

While the induction generator is synchronised to the power system its terminal voltage and frequency are fixed by the system. If an induction generator becomes disconnected from the system, then both its voltage and its frequency lose synchronism. This condition, sometimes known as "islanding", can cause serious problems by generating power outside the prescribed limits for voltage or/and frequency and also for continuing to deliver power to a system fault.

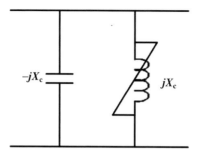

Fig. 8.8 Equivalent circuit of a disconnected induction generator.

When a generator becomes isolated from the system network, it tends to accelerate as load is removed. The increase of speed is accompanied by a corresponding increase of frequency of the generated voltages. A loss of system connection of the equivalent circuit Fig. 6.17 can be reduced to that of Fig. 8.8, representing only the dominant impedances of the generator magnetic circuit jX_m and the external power factor correction impedance $-jX_c$. If the connected capacitance of the power factor correction equipment is sufficient to equal the no-load reactive power requirement of the generator, there arises a possibility of self-excitation. This resonant condition, whereby energy oscillates between the generator magnetic field and the power factor correction capacitors, can give rise to excessive voltage levels that may become sufficiently high to damage connected equipment. It is customary for compensated wind turbines to have fast-acting relay protection to detect and shut down the turbine if it should become disconnected.

Calculation of the resonant frequency and the terminal voltage during the condition of self-excitation is complicated by magnetic saturation of the wind generator. This is reflected in the equivalent circuit of Fig. 8.8 by the fact that machine reactance X_m is a nonlinear function of the magnetising current.

References

1. Walker, J. F., N. Jenkins, *Wind Energy Technology*, John Wiley & Sons, Chichester, England, 1997.

2. Burton, T., D. Sharpe, N. Jenkins, E. Bossanyi, *Wind Energy Handbook*, John Wiley & Sons, Chichester, England, 2001.
3. The Minnesota Public Utilities Commission, Final Report — 2006 Minnesota Wind Integration Study, Minneapolis, the USA, 30 November, 2006.
4. Wikipedia, 13 August 2010. http://en.wikipedia.org/wiki/Installed_wind_power_capacity.
5. van Werven, M. J. N., L. W. M. Beurskens, J. T. G. Pierik, Integrating Wind Power in EU Electricity Systems (Economic and Technical Issues), European Community GreenNet Project Report, Contract Number NNE5-2001-660.
6. Milligan, M. R., Modelling Utility-Scale Wind Power Plants, Part 2: Capacity Credit, National Renewable Energy Laboratory, Golden, Colorado, the USA, Report NREL/TP-500-29701, March 2002.
7. EnerNex Corporation, Knoxville, Tenn, the USA and Wind Logics Inc., St Paul, Minn, the USA, Wind Integration Study — Final Report, Prepared for Xcel Energy and the Minnesota Department of Commerce, Minneapolis, Minn, the USA, September 2004.
8. Ford, R., D. Milborrow, Integrating Renewables, British Wind Energy Association, 9 pages, available at http://www.bwea.com/pdf/RAEIntegrationfinal.pdf, February 2005.
9. Bossanyi, E., Z. Saad-Saoud, N. Jenkins, Prediction of Flicker Produced by Wind Turbines, *Wind Energy*, **1**, September 1998.
10. Copper Development Association, Earthing Practice, CDA Publication No. 199, England.
11. Cotton, I., *et al.*, Lightning Protection for Wind Turbines, Interrational Conference on Lightning Protection, pp. 848–853, 2000.
12. Manwell, J. F., J. G. McGowan, A. L. Rogers, *Wind Energy Explained — Theory, Design and Application*, John Wiley & Sons, Chichester, England, 2002.
13. Lubosny, Z., Wind Turbine Operation in Electric Power Systems, Springer-Verlag, Germany 2003.
14. Williams, B. W., *Power Electronics*, Second Edition, The Macmillan Press, Hampshire, England, 1992.
15. IEEE Recommended Practices and Requirements for Harmonic Control in Electrical Power Systems, ANSI/IEEE Standard 519-1992, IEEE, New York, the USA, 1992.

CHAPTER 9

Environmental Aspects of Wind Energy

The British Wind Energy Association (BWEA) has issued a set of best practice guidelines, including environmental features, of wind turbine and wind farm operation.[1] Other countries operate corresponding systems and these also normally take the form of voluntary advice rather than legislated prescriptions. The BWEA guidelines topics are summarised in Table 9.1, with comments in a separate column.[2,3] Some environmental features listed in the table are discussed in greater detail in further subsections of this chapter.

9.1 Reduction of Emissions

One of the great advantages from the use of wind energy to provide electricity is the consequent reduction of the use of fossil fuel. In addition to the saving of the fuel, the elimination of fossil fuel combustion represents a corresponding reduction in the emission of pollutants. The most polluting of the fossil fuels is coal.

9.1.1 *World consumption of coal*

Figures for the consumption of coal over the period 1997–2007 are given in Table 9.2.[4] Consumption reduced in some of the European countries due

Table 9.1 BWEA environmental statement regarding wind energy practice.[1]

Topic	Comment
Policy framework	Any proposal is tested for compliance with regional and national development plans.
Site selection	Justification for the choice of site.
Designated areas	Does a proposed project affect any areas considered to be of special interest (e.g., national parks)?
Visual and landscape assessments	Incorporates zones of visual influence from which the wind farm will be visible. A highly subjective feature.
Noise assessment	Also highly subjective. Proximity of the nearest dwellings, both upwind and downwind.
Ecological assessment	Impact on local flora and fauna.
Archaeological and historical assessments	To avoid interference with any sites of archaeological and/or historical interest.
Hydrological assessment	Any impact of a proposal on water courses and water supplies.
Interference with telecommunication systems	To ensure that there is no interference with TV signals or with communication (microwave systems or radar systems).
Aircraft safety	Both civil and military low-flying aeroplanes must be free from hazard.
Site safety	The structural integrity of the turbines, local road safety, and shadow flicker effects.
Traffic management	Any modification requirement for public highways plus the impact of access roads.
Electrical connection	Environmental impact of overhead transmission lines and of electrical sub-stations.
Economic effects on the local community	Any permanent or temporary jobs that may be created.
Environmental effects in general	Notably the reduction of greenhouse gas emissions.
Tourism and recreational effects	Impact on local tourism and on recreational facilities.
Decommissioning	The removal of all above ground equipment and surface restoration at the end of its useful life.
Mitigating measures	Methods by which any adverse environmental effects may be mitigated.
Non-technical summary	To make accessible to all readers the results of environmental assessments.

Table 9.2 Coal consumption (extracted from Ref. [4]) (million tons of oil equivalent).

Country	1997	2007	% Change
The USA	540.4	573.7	+6.16
Canada	27.9	30.4	+8.96
France	13.4	12.0	−10.40
Germany	86.8	86.0	−0.09
Italy	11.0	17.5	+59.10
The Netherlands	9.5	8.8	−7.40
Russian Federation	107.1	94.5	−11.80
Spain	17.7	20.1	+11.90
The UK	39.6	39.2	−1.01
China	700.2	1,311.4	+87.30
India	135.9	208	+53.10
Indonesia	8.2	27.8	+239.00
Japan	89.8	125.3	+39.50
South Korea	21.7	41.1	+90.80
World	**2,317.7**	**3,177.5**	**+37.10**

to the increased use of natural gas but the world consumption increased by 37.1% over the period. In particular, the industrialising countries of the Pacific Rim show marked increases of consumption. Over the same period, the consumption of natural gas of the world has increased from 2026.4 to 2637.7 million tons of oil equivalent, representing an increase of 30.2% and the consumption of oil increased from 3433.3 to 3952.8 million tons, being an increase of 15.1%.

In the UK, coal is still the prime fuel burned in the production of electricity. In Western Europe and North America, most of the coal consumption is now used for generating electricity. Large power stations burn pulverised coal in a suspension of fine particles. A 1,000-MW$_e$ coal-fired plant burns about 2.5 million tons of coal and generates roughly 6.5 billion kWh of electrical energy per year.

9.1.2 *Open coal fires*[5]

Until the 1950s, coal was the prime domestic and industrial fuel in the UK and was mostly burned on open fires. The smoke and gaseous effluent from millions of fires was expelled directly into the atmosphere above residential areas. Some degree of effluent cleansing or filtration was carried out by some industries, including the electricity-generating industry, but the overall effect of open-fire coal burning, mostly residential, was massive air pollution. This was evidenced in the smoke blackening of buildings stone and brick and in the incidence of thick fogs in the industrial cities. A great fog in London in 1954 was a mixture of natural fog and man-made smoke and was called "smog". This lasted several days, brought transportation to a standstill and is thought to have been directly responsible for hundreds of accelerated deaths due to lung disease.

Apart from the domestic inconvenience, the use of coal-burning open fires or coke-burning stoves is dirty, unhealthy, and grossly inefficient. Most of the heat goes up the chimney and is wasted. The number of victims, including fatalities, of lung diseases created by or compounded by 200 years of open coal fires cannot be calculated but is probably enormous in the UK alone. A succession of legislation, the "Clean Air Acts", has largely banned open-fire coal burning in UK cities, which are now described as "smokeless".

9.2 Effluents due to Coal Burning[5]

The three main pollutants from coal-burning plant flue gases are sulphur dioxide (SO_2) and nitrogen oxides that produce "acid rain", plus particulates.

9.2.1 *Sulphur oxides*

The most important pollutants released by coal burning are sulphur products. "Inorganic" sulphur (sometimes called "mineral" sulphur) is physically distinct from but attached to the coal and can be largely removed by washing prior to combustion. "Organic" sulphur is chemically combined with the coal material and cannot be washed out. Emissions of sulphur dioxide or SO_2 are measured in terms of pounds (or kilograms) per million BTU burned.

Sulphur may be removed from the combustion gases by chemical "scrubbing". The sulphur-oxide-laden exhaust gases are brought into contact with a scrubbing agent, such as lime or limestone, to produce a chemical reaction that removes the sulphur. The stream of effluent from the absorbent scrubbers has its water evaporated off, producing a sludge that has to be disposed of as solid waste. For example, a unit burning 2% sulphur coal produces about 200 lb of sludge (dry weight) per ton of coal burned. A power plant of (say) 500 MW$_e$ would create a 560-acre sludge disposal area 40 ft deep, over its lifetime. More modern scrubbing agents such as sodium or recyclable metal hydroxides neutralise the sulphur oxides and also reduce the scaling of the pipes.[5]

9.2.2 Nitrogen oxides

During the combustion of coal, nitrogen is drawn both from the coal and from the air in which it burns. Nitrogen oxide formation depends on the flame temperature, the time of combustion, the excess air present, and the rate of cooling. To lower nitrogen oxide emissions, it is necessary to modify the combustion processes. Design arrangements are effective if they involve lower combustion temperatures, restrictions of air intake, recirculation of flue gases, and injection of water into the fire-box.[6] Motor vehicle exhaust gases are a bigger source of nitrogen oxides than power station flue gases in the UK. For this reason, there is now UK legislation requiring all new cars to be fitted with exhaust gas filters. In the USA, about one-half the nitrous oxide emissions in 1999 were due to the nitrogen fertilisation of soils in the agricultural sector.[7]

Acid precipitation or "acid rain" is caused by SO_2 and nitrogen oxides mixing chemically with water vapour in the atmosphere. These emissions are mainly derived from coal and oil combustion, which is dispersed through tall chimneys to prevent high concentrations at ground level. Due to air motion, increases in the acidity of local rainfall may occur hundreds of miles downwind from the site of the pollution. The emitted gases may be oxidised to sulphates and other chemical changes may occur, influenced by varying conditions of temperature, humidity, and solar irradiation. Also, the chemical nature of the gases may be affected by substances originating in the territories over which they travel.

There is evidence that pollution from power plants in the UK is blown from west to east and deposits, in the form of acid rain, onto locations in Norway and Sweden. The acidity of rain can accelerate the leaching of ground toxic materials such as aluminium and mercury into water courses. There is also evidence of damage to forests and fish stocks, especially in the areas with acid soils.

The problem of acid rain has to be approached based on international collaboration. Technical aspects of the problem of acid rain can only be tackled at source. There is no form of prevention once the acidity is created.

9.2.3 *Particulates*

Flue gases from coal-burning operations contain small particles of solid materials, mainly carbon. There is a range of particle sizes down to less than 1 mm, and the suspensions of smaller particles can be ingested or inhaled by respiration. Particulates may travel hundreds of miles from their point of origin, accompanying emissions of gas from the same source. The chemical nature of particulates can change in transit. Both increases and decreases of toxicity have been noted. Also, during transit, the emitted gases may form aerosols (fine suspensions) through condensation and coagulation. These may react with other fine suspensions, leading to increases of particulate size with changed physiological effects, if ingested or inhaled into human bodies.

In order to satisfy stringent environmental regulations, chimney effluents are controlled by one or more of four filtration processes: mechanical collectors, electrostatic precipitators, scrubbers, and fabric baghouses. All the methods are more effective in capturing the larger size particles. Sometimes, two types are used in series, with a cheap and relatively inefficient first stage acting to reduce the loading on a more efficient and more expensive second stage.[5]

Mechanical collectors use gravity, inertia, or centrifugal forces to separate (mainly) heavier particles from the gas. The simplest form consists of enlarged chambers in the gas stream that slows down the flow rate, enabling the heavier particles to travel the outer walls and drop to the bottom of the chamber.

Electrostatic precipitators are used by the UK electricity generation industry. The flue gases are passed between a high-voltage electrode and a grounded collection plate. The ionised particles move towards the grounded plate, where they are collected and removed. As much as 99.9% (by weight) of the particles can be removed but the precipitation works best on the heavier particles. With low sulphur coals, the particles tend to be electrically more resistive. This makes electrostatic site precipitation more expensive because the precipitators may then need to operate in a hotter part of the flue gas, where the higher temperature reduces the resistivity of the carbon particles. Alternatively, bigger precipitators may be required.

Wet scrubbers are sometimes used to wash solid particles from the gas stream using water. This is essentially a physical scrubbing process that is different from the chemical scrubbing of sulphur oxides, described above.

For the filtration of the finer particles, the most effective method is to use fabric filter baghouses. The flue gas is forced through fine filters, effecting filtration but causing a pressure drop. This increases the operating costs. The method is widely used for general purpose industrial applications. High temperature and corrosive gases due to coal combustion in power utility boilers pose particular problems for filter methods.[8]

9.2.4 Carbon dioxide

The burning of coal also releases carbon dioxide (CO_2) gas and thereby contributes to the possibility of global warming due to the accumulation of so-called "greenhouse" gases.

Table 9.3 gives figures for past and projected future emissions of CO_2 due to the use of coal.[8] CO_2 emissions due to coal are found to be smaller than those due to oil in the OECD countries and the Middle East but much larger in China. There is a close correlation between the amount of coal consumed (Table 9.2) and the consequent amount of CO_2 emitted (Table 9.3).

There is now a broad scientific consensus view that global warming is due to the buildup of CO_2 gas in the earth's stratosphere. Moreover, the view regards the buildup of CO_2 as due, at least partly, to the combustion of fossil fuels, mainly coal. There is some scientific dissent from the consensus view, attributing climate change entirely to natural climatic behaviour.[9]

9.3 Wind Turbine Noise[2,3,10]

Two types of noise are generated by wind turbine operation: aerodynamic noise from the rotating blades and mechanical noise from the rotating machinery. Wind turbines are usually located in the areas of low background noise that tends to exacerbate their own noise effect.

Table 9.3 World CO_2 emissions from coal use by region, reference case, 1990–2005 (million metric tons CO_2).

Region/country	History				Projections				Average annual percentage change 2005–2025
	1990	2001	2002	2010	2015	2020	2025		

Region/country	1990	2001	2002	2010	2015	2020	2025	Average annual percentage change 2005–2025
Mature Market Economies								
North America.....	1,923	2,214	2,247	2,537	2,616	2,774	3,110	1.4
The United States[a]	1,784	2,042	2,070	2,335	2,407	2,561	2,858	1.4
Canada....	123	148	154	166	169	172	210	1.4
Mexico....	15	24	23	36	39	41	42	1.4
Western Europe.....	1,160	849	825	783	740	699	661	−1.0
Mature Market Asia.....	383	593	622	675	671	690	689	0.4
Japan.....	245	364	375	386	380	376	372	0.0
Australia/New Zealand	138	229	247	289	291	314	317	1.1
Total Mature Market.....	3,465	3,656	3,694	3,995	4,026	4,162	4,459	0.8
Transitional Economies								
Former Soviet Union.....	1,222	667	671	761	781	796	812	0.8
Russia.....	635	408	390	456	473	471	465	0.8
Other FSU.....	587	260	281	304	308	325	347	0.9
Eastern Europe.....	685	409	397	408	412	414	419	0.2
Total Transitional.....	1,907	1,076	1,068	1,169	1,193	1,210	1,231	0.6

(Continued)

Table 9.3 (Continued)

Emerging Economies								
Emerging Asia.....	2,607	3,679	3,831	5,784	6,724	7,488	8,024	3.3
China.....	1,886	2,472	2,582	4,181	4,911	5,514	5,887	3.6
India.....	394	684	698	903	1,019	1,119	1,222	2.5
South Korea.....	90	152	161	196	235	263	286	2.5
Other Asia.....	237	371	389	504	559	592	629	2.1
Middle East.....	**72**	**110**	**100**	**138**	**140**	**140**	**139**	**1.4**
Africa.....	**271**	**340**	**3368**	**421**	**453**	**484**	**486**	**1.6**
Central & South America.....	**54**	**66**	**73**	**98**	**106**	**114**	**120**	**2.2**
Brazil.....	34	40	41	54	59	66	74	2.6
Other Central/South America.....	20	26	32	44	47	48	46	1.6
Total Emerging.....	**3,003**	**4,196**	**4,342**	**6,441**	**7,423**	**8,227**	**8,768**	**3.1**
Total World.....	**8,375**	**8,928**	**9,105**	**11,604**	**12,642**	**13,600**	**14,458**	**2.0**

[a]Includes the 50 States and the District of Columbia.

Sources: **History**: Energy Information Administration (EIA), *International Energy Annual 2002*, DOE/EIA-0219 (2002) (Washington, DC, March 2004), website www.eie.doe.gov/iea/. **Projections**: EIA, *Annual Energy Outlook 2005*, DOE/EIA-0383 (2005)(Washington, DC, February 2005), October futures case, AEO2005 National Energy Modelling System, run CF2005.D111104A, website www.eia.doe.gov/oiaf/aeo/; and System for the Analysis of Global Energy Markets (2005).

The aerodynamic noise consists of the swishing sound of the rotor blades scything through the air. It is similar to the stirring of tree branches during a brisk wind. This is likely to be detectable only for limited periods during low wind speeds. At higher wind speeds, the ambient wind noise from trees and buildings is likely to mask the wind noise of turbine blades.

The number and layout of turbines on a particular wind-farm site may be limited by noise considerations.

The response of people to noise from wind turbines is highly subjective. In a country area, for example, the ambient noise is likely to be lower than in an urban area. What is acceptable in an urban area may seem unreasonable intrusive in a more remote location.

9.3.1 Measurement of wind turbine aerodynamic noise

Noise can be quantified in precise terms; the sound power level of the source and the sound pressure level at a given location can both be quantified on logarithmic scales and are measured in the unit decibels (dB). This is a standard unit representing signal gain or attenuation that is widely used in telecommunications and wireless engineering. The sound level of a source L_W can be defined as:

$$L_W = 10 \log_{10} \left(\frac{W}{W_o} \right). \qquad (9.1)$$

In Eq. (9.1), W is the sound level of the source and W_o is a reference sound level, usually taken, in electrical terms, to be 1 picowatt or 10^{-12} watt. If the sound level doubles, the $L_W = 10 \log_{10} 2 = 10 \times 0.301 = 3$ dB. The sound level of a large modern wind turbine at its normal rotational speed is of the order 100 dB. From Eq. (9.1), a level of 100 dB corresponds to $\log_{10}(W/W_o) = 10$ or $(W/W_o) = 1,000,000$. In other words 10^{-6} watt or 1 microwatt.

The sound pressure level L_P can similarly be expressed in dB from the defining equation:

$$L_P = 10 \log_{10} \left(\frac{P}{P_o} \right), \qquad (9.2)$$

where P is the rms sound pressure level of the source (i.e. the wind turbine) and P_o is a reference sound pressure level, often taken to be 20 micropascal

(20 μPa) or 20×10^{-6} Pa. A typical value of sound pressure 350-m distance from a wind farm has been measured of the order 35–45 dB.[2]

The use of the definitions of Eq. (9.1) or (9.2) permits sound changes of high order, found widely in practice, to be expressed in straightforward linear terms on a dB scale.

The branch of technology called control systems engineering, incorporating cybernetics and robotics, uses a terminology to define system gain or attenuation in terms of frequency response ratio f/f_o. The defining equation is similar to Eq. (9.2) except that a multiplier 20 is used, rather than 10.

In the discipline of music, for example, a doubling of the sound frequency is called an octave. From this arises use of the terms 6 dB/octave and 20 dB/decade. These terms are extensively used to define the performance of electrical and electronic devices and systems.

The total sound pressure level at a location due to several wind turbines can be obtained by the addition of terms, such as that of Eq. (9.2).

$$L_{\text{total}} = 10 \log_{10}\left(\frac{P_1}{P_o}\right) + 10 \log_{10}\left(\frac{P_2}{P_o}\right) + \cdots$$
$$= 10 \log_{10}\left(10^{\frac{P_1}{10}} + 10^{\frac{P_2}{10}} + \cdots\right). \qquad (9.3)$$

For a total of N noise sources, the total sound pressure can be generalised as:

$$L_{\text{total}} = 10 \log_{10} \sum_{i=1}^{N} 10^{\frac{P_i}{10}}. \qquad (9.4)$$

There are no universally agreed international noise standards or regulations. Many countries have regulations that define upper limits for the level of industrial noise to which people may be exposed. These limits are frequently exceeded in Western countries at concerts of popular music. Some countries apply different (lower) recommendations of noise level at night, compared with the daytime. Regulations pertaining to residential areas are of the same order as those for rural areas. Recommended maximum noise levels for three European countries are given in Table 9.4. The order of noise levels recommended, being in the range 30–60 dB, are typical of the normal noise levels in residential premises. The threshold of pain for the human ear is about 200 Pa, which corresponds to a sound pressure level of 110 dB.[10]

Table 9.4 Noise limits of equivalent sound pressure levels (dB).[11]

Country	Commercial	Mixed	Residential	Rural
Denmark	—	—	40	45
Germany				
day	65	60	55	50
night	50	45	40	35
The Netherlands				
day	—	50	45	40
night	—	40	35	30

9.3.2 *Mechanical noise*

In addition to aerodynamic noise from the motion of the blades in the air, a wind turbine creates mechanical noise from the rotation of the gearbox and generator, plus auxiliary equipment. This noise is of comparable order with that due to any rotating machinery equipment but is more noticeable and perceptible because a wind turbine is located in the open and often remote from other equipment. The noise created by rotating machines, at constant rotational speed is of narrow waveband and sometimes of singular frequency but it may have a broadband component. The sound transmission path from a wind turbine may be airborne or structure borne from the tower, hub, or rotor, which can act as loudspeakers.

Some measured data are given in Table 9.5 from experiments at 115 m distance downwind from a 2-MW turbine. The main sources of mechanical noise that created airborne noise and/or caused radiated noise from the nacelle surfaces and the machinery enclosure are listed.[10] It is seen that the gearbox alone created a structure borne noise that is approaching the level of the aerodynamic noise. The total predicted sound power noise in this case is given as 102.2 dB, which is the order of sound pressure created by heavy street traffic in a city and exceeds the aerodynamic noise by 3 dB. The intensity of the sound pressure received at any point in air varies (almost) with the inverse square of the distance from the transmitter. Therefore, the best mechanism to reduce the effect of any acoustic sound is to move further away from the transmitter (i.e., the wind turbine). This precept is closely observed in the siting of wind turbines and wind farms with regards to residential areas. In the UK, for example, there is accepted guidance on the

Table 9.5 Noise levels at 115 m downwind for a 2-MW turbine[12] (adapted from a diagram in Ref. [10]).

Source of noise		Measured value of sound pressure (dB)
Aerodynamic		99.2
Mechanical (airborne)	Generator	87.2
	Gearbox	84.2
	Auxiliaries	76.2 $L_W = 102.2$ dB (estimate)
Mechanical (structure borne)	Gearbox	97.2
	Hub	89.2
	Blades	91.2
	Tower	71.2

measuring and predicting of noise levels from wind farms.[13] Other countries produce corresponding official advice.[14]

9.4 Electromagnetic Interference from Wind Turbines[15]

The operation of a wind turbine can cause interference with the signals of radio, television, and microwave systems in two ways.

1. The turbine operation can create radio frequency noise (30–200 MHz), especially in the presence of the power electronic switching devices used in variable speed systems.
2. The action of the turbine, especially the blade rotation, can cause the interruption and scattering of electromagnetic signals that strike the structure.

9.4.1 *Electromagnetic interference radiated from wind turbines*

The hilltop sites used by wind turbines are also good propagation sites for telecommunication transmitters and relays. Although the generator and its control gear can produce radio frequency emissions, these can be suppressed and screened by standard techniques. For example, a radio signal cannot penetrate thickness of metal or a well-designed cage mesh. Metal tubular towers have a screening action, as well as being more visually appealing than lattice structures, for many people. The variable frequency inverter systems described in Chap. 7 all operate by the rapid action of semiconductor switches. This creates radio frequency interference by both transmission along the system conductors and propagation through the air. The airborne signals would be a serious problem with local radio communication but can readily be suppressed at the source. Overall, the electromagnetic emissions from wind turbines do not constitute a serious problem.

9.4.2 Electromagnetic interference effects due to the rotating blades

The rotation of the propeller blades of a wind turbine causes a scattering effect on a radio signal that strikes it. Some component of the radio beam may be reflected back along its transmission path. Some proportion of the incident signal may be deflected along a different path. A radio receiver may receive two signals from the same source simultaneously, one signal being the original transmission together with a second signal that has become distorted or time delayed by scattering.

When the wind turbine is between the transmitter and the receiver (Fig. 9.1(a)), the effect is known as "forward scattering" or front-scatter. For TV signals (300 MHz–3 GHz), there can be a jittering of the picture at the blade passing frequency of 1–2 Hz.[10] When the turbine is located behind the receiver (Fig. 9.1(b)), the effect is known as "back scattering" or back-scatter and can cause double or ghost images.

Microwave communication systems (1–30 GHz) such as those used for air-traffic control are vulnerable to interference from signal scattering caused by wind turbines. These communication systems are of longer standing than the more recent technology of wind-powered electricity generation. In the planning and preparation stages for the siting of a wind farm, it is necessary to be aware of the beaming and lines of sight of any communication system in the area. In the USA, for example, existing rules from the Federal Aviation Authority prevent any structure the size of a wind turbine being erected within one kilometre of an omni-directional range-finding VHF station for aircraft navigation and landing.[10]

A much more detailed and mathematical consideration of these issues is given in Ref. 15.

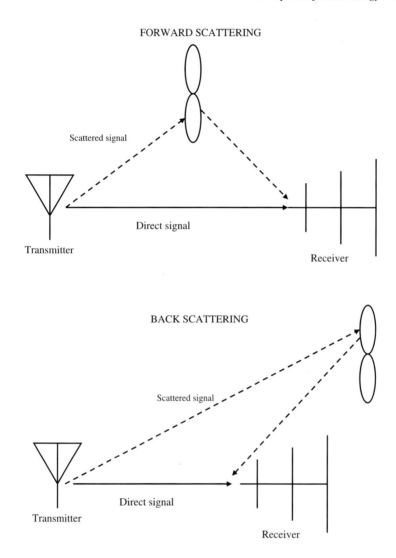

Fig. 9.1   Interference mechanisms of wind turbines with radio systems.

9.5 Effect of a Wind Turbine on Wildlife

The operation of a wind turbine does not appear to have any effect on ground animals. The land on which a turbine or wind farm stands is still available for agricultural use. Cattle and sheep can graze right up to the tower structure and do not appear to be affected by the sight and sound of the blade rotation.

Bird populations can be affected by the building and operation of wind farms. During construction, there may be loss of bird habitat. Operation of a wind farm may cause changes to bird foraging habits, disturbance of breeding and nesting behaviour, and an alteration of migration habits. Bird kills due to collision with the rotating blades have been widely reported across the world. Large rated wind turbines operating at lower wind speeds do not usually claim as many victims as the smaller and faster turbines. Bird deaths also occur due to collisions with fixed structures such as bridges, radio transmission towers, and power transmission lines.

Many studies have been made to estimate bird mortality due to wind farms. Particular problems arise where large numbers of birds congregate or migrate. Mountain passes are frequently windy and provide a good wind channel for turbines but may also be a preferred route for migratory birds. Major bird migration corridors and areas of high bird concentration should usually be avoided in siting wind facilities.[10]

In the UK, the wind farm at Blyth Harbour in the north of England consists of nine 300 kW wind turbines positioned along the breakwater. This particular location has been identified as having 110 varieties of local birds — the highest bird density of any UK wind-farm site.

Over a three-year period, there were 31 collision victims identified, mainly among the three species of gulls. This was considered to be not a significant proportion of the local bird populations. The studies taken on the coastal wind farms in Holland and Denmark, together with UK experience, suggest that most species had adapted to the turbine presence. There was no significant impact on habitat loss. Although displacement of the cormorants has occurred during construction at one site, the cormorant population returned to its previous haunts after construction was completed.[3]

A 1994/95 study of the Bryn Titli wind farm in Wales, which is close to breeding communities of buzzard, peregrine, red grouse, snipe, curlew and raven, contains 22 turbines of 450 kW rating, suggested that there were no bird strikes at all in the period.[3]

Any bird deaths at all are to be regretted. But an extensive review of both American and European literature of the impact of wind turbines on birds concludes that, of the mainly coast locations studied, the risk of death by collision with wind turbines is minor on the population level of local species.[3]

9.6 Visual Impact of Wind Turbines

9.6.1 *Individual response*

The reaction of people to the sight of a wind turbine or a wind farm is highly subjective and individual. A particular wind farm may appear visually intrusive to some people but not to others. Because of their height and location, wind farms are conspicuous and can be seen from a distance. Most people will only ever see a wind farm from a distance rather than close up. The visual appearance, and especially the distant aspect, must be considered at an early stage of the design process.

Opinions regarding the visual impact of wind farms are based on individual values and judgements and are influenced by issues such as:[16]

- The value that a viewer places on the preservation of the proposed site, its surrounding area, and its social and historic context.
- The value that a person places on the clean production of electricity and reduction of pollution.
- A person's familiarity with the technology of wind-powered electricity and the alternatives.
- A person's interest in and awareness of energy supply and demand.

An overall view might be based on balancing these issues. There can be no single right answer. In Europe and in North America, there is much governmental advice and many references and handbooks contain design guidelines for minimising the visual environmental impact of wind turbines and wind farms.[17]

9.6.2 *Shadow flicker*[3,10]

The moving blades of a wind turbine rotor can cast moving shadows that cause a flickering effect. Similarly, sunlight may be reflected from the blades creating a flashing effect. Both effects are stroboscopic. The shadow flicker can cause variations of light in the range 2.5–3 Hz, which can affect some sufferers from epilepsy. Large three-bladed turbines usually operate at rotational speeds lower than 35 rpm giving blade-passing frequencies less than 1.75 Hz.

Because of the latitude and low angle of the sun in the winter sky, this phenomenon has been experienced in Northern Europe but not in the USA. The experience is exacerbated in Europe because of the near spacing between dwelling houses and the turbines. In Denmark, the common guideline is to have a minimum distance of six to eight rotor diameters between a turbine and its closest neighbour. In the UK, a minimum spacing of 10 rotor diameters is recommended, which is consistent with the spacing required by noise constraints.

9.7 Safety Aspects of Wind-Turbine Operation[10]

Wind-turbine sites are usually available for visits from members of the public. Visitors may be exposed to certain safety hazards because of the operation of moving machinery. A turbine nose cone or nacelle cover may be blown off in a high wind. Blade failure may occur resulting in a rotor blade being thrown off. Most fragments shed by a turbine have been found within 100–200 m of the tower, in the rare incidents that have been recorded.

The most critical components on a wind turbine in terms of fatigue loading are the blades and the gearbox. At present, the majority of blades, made of glass-reinforced plastic, are designed to last for 20 years. Accelerated fatigue tests and two years of testing are standard before new designs are included in products.[18]

In some weather conditions, there can be a buildup of ice on the rotor blades. If this ice falls to the ground or is thrown off, it can obviously be hazardous to people in the vicinity. The installation and maintenance of wind turbines possess engineering and construction problems to the operating personnel. There have been cases of accident, sometimes fatal, during the buildup of the industry. No figures appear to be available to compare industrial accidents in the wind-energy industry with other energy-producing operations.

References

1. BWEA, Best Practice Guidelines for Wind Energy Development, British Wind Energy Association (BWEA), London, England, available at www.bwea.com, 1994.

2. Walker, J. F., N. Jenkins, *Wind Energy Technology*, John Wiley & Sons, Chichester, England, 1997.
3. Burton, T., D. Sharpe, N. Jenkins, E. Bossanyi, *Wind Energy Handbook*, John Wiley & Sons, Chichester, England, 2001.
4. BP Statistical Review of World Energy, British Petroleum Company Inc., London, England, June 2008.
5. Shepherd, W., D. W. Shepherd, *Energy Studies*, Second Edition, Chapter 4, Coal, Imperial College Press, London, England, 2003.
6. Energy in Transition 1985–2010, Final Report of the Committee on Nuclear and Alternative, Energy Systems, Chapter 4, National Research Council, Washington, D.C., the USA, 1980.
7. Emissions of Greenhouse Gases in the USA, 1999 — Executive Summary, US Energy Information Administration Report EIAIDOE — 0593 (99), US Department of Energy, Washington, D.C., the USA, October 2000.
8. International Energy Outlook 2005, US Energy Information Administration Report DOE/EIA — 0484 (2005), Washington, D.C., the USA, July 2005.
9. Robinson, A. B., N. E. Robinson, W. Soon, Environmental Effects of Increased Atmospheric Carbon Dioxide, *Journal of American Physicians and Surgeons*, **12**, 79–90, 2007.
10. Manwell, J. F., J. G. McGowan, A. L. Rogers, *Wind Energy Explained*, John Wiley & Sons Ltd., Chichester, England, 2002.
11. Gipe, P., *Wind Energy Comes of Age*, John Wiley & Sons Inc., New York, the USA, 1995.
12. Wagner, S., R. Bareib, G. Guidati, *Wind Turbine Noise*, Springer Verlag, Berlin, Germany, 1996.
13. The Assessment and Rating of Noise from Wind Turbines, Report ETSU-R-97, UK Department of Trade and Industry (DTI), London, England, 1996.
14. National Wind Coordinating Committee (NWCC), Permitting of Wind Energy Facilities: A Handbook, RESOLVE, Washington, D.C., the USA, 1998.
15. Sengupta, D. L., T. B. A. Senior, *Wind Energy Technology*, Chapter 9, Electromagnetic Interference from Wind Turbines, ASME (American Society of Mechanical Engineers), New York, the USA, 1994.
16. Wind Power: Environmental and Safety Issues, Wind Energy Fact Sheet 4, Renewable Energy Enquiries Bureau, Department of Trade and Industry, London, England, August 2001.
17. Stanton, C., *The Visual Impact and Design of Wind Farms in the Landscape*, Wind Energy Conversion, pp. 249–255, British Wind Energy Association (BWEA), London, England, 1994.
18. Connellan, M., *Spinning to Destruction*, The Guardian, London, England, 4 September 2004, available at http://www.guardian.co.uk/technology/208/sep/04(energy.engineering).

CHAPTER 10

Economic Aspects of Wind Power

In order to be economically worthwhile, the income produced from the wind-generated electricity has to exceed the total cost of installing and operating the generating plant. The manufacturing and selling of electricity is the same, economically, as any other business. Both the "income" and "expenditure" sides of the balance sheet involve the detailed consideration of several important variables and features that are discussed in appropriate sections below.

10.1 Investment Aspects of Wind-Powered Electricity Generation

The cost of a wind-energy system has two basic components; installation costs and operation costs. The factors that influence the costs of wind-powered electricity generation includes:

- the costs of the turbines and generators,
- the costs of the turbine site, construction, and grid connection,
- the operation and maintenance costs during the system lifetime,
- the turbine lifetime and depreciation rate,

- the business costs associated in the financing of the building and operation,
- the wind regime at the site,
- the energy capture efficiency of the turbines, and
- the availability for sale of the generated electricity.

Wind-energy project capital costs, reported by the International Energy Agency, show substantial variation among countries due to the factors such as market, structures, site characteristics, and planning regulation. The installation cost can be expressed as a function of the rated electrical capacity.

At the latter part of the 1990s, the total wind project capital costs varied between approximately US$900 per kW and US$1600 per kW in different countries.[1] Although the cost varies among different countries, the trend everywhere is the same—wind energy is getting cheaper. The cost of turbines is getting cheaper/kW rating as the technology improves and the components can be made more economically. The productivity of the newer designs is also better so that more electricity is produced from more cost-effective turbines. There is a trend towards larger machines that reduces infrastructure costs as fewer turbines are then needed for the same output.[2]

More recent figures for European investment costs for onshore wind farms are €1150/kW ($1300/kW) at an average wind speed of 9.5 m/s.[3] The costs for offshore wind generation are significantly higher both for the building and for the maintenance aspects. These can amount to at least 50% greater than onshore installation of comparable rating and are sometimes twice as much.

10.1.1 *Costs of the turbines and generators*

The capital cost of the turbine/generator units is by far the most expensive item in the list of components for setting up a wind farm. A comparison between the USA and the Netherlands in 1993 (Table 10.1) gave figures of about 75% for the proportional cost of the turbine. In 1997, a wind farm capital cost breakdown (Table 10.2) gave a figure of 64% for the cost of the turbine. A more recent (2007) breakdown of costs once again repeated the figure of 64% for the turbine.[2]

The average size of turbines sold in the Danish export market increased from roughly 50 kW in 1985 to 500 kW in 1996. By 1998, the best-selling

Table 10.1 Wind farm capital cost breakdown (%) for the USA and the Netherlands, 1993.[4]

Components	The United States	The Netherlands
Wind turbine	74.4	76.8
Engineering and installation	6.7	3.4
Tower and foundation	11.1	6.6
Grid connection	6.9	11.8
Roads	0.9	1.4

turbine was rated at 600 kW and this constituted 80% of the cost of an onshore wind farm installation. The general investment costs per kilowatt declined by more than five per cent per year from 1989 to 1996.[1] A more detailed cost for a breakdown, for a 50-turbine wind farm, is given in Table 10.3.[6,7] The generator is of relatively low cost compared with the transmission and drive train that includes the three-stage (usually) or two-stage gearbox, for large machines.

10.1.2 Costs of the turbine site, construction, and grid connection

The building of a wind farm site usually involves major construction features such as the construction of access and service roads and the construction of the foundations for the turbine towers, the associated transformer, and the switching stations. Buildings are required to house the instrumentation, control, and telecommunications systems. When personnel live on site, it is necessary to provide appropriate living accommodation.

For onshore locations, the land site has to be acquired by purchase or leasing. Parts of the site may be suitable for rental to local farmers for animal grazing once the wind farm is in operation. The electrical infrastructure and grid connection amount to a significant sum. In Table 10.2, for example, these items account for 14% of the installed cost. For the large installation described in Table 10.3, the "electrical items", excluding the generator, amount to 15.5% of the total installed cost. The construction costs for wind-powered electric generators are considerably higher than those of fossil-fuel plants on a per megawatt basis. For example, in 2008, the construction costs

Table 10.2 Wind farm capital cost breakdown for a UK 5-MW Installation.[5]

Components	Percentage of total cost
Wind turbine	64
Civil works	13
Electrical infrastructure	8
Grid connection	6
Project management	1
Installation	1
Insurance	1
Legal/development costs	3
Bank fees	1
Interest during construction	2

Table 10.3 Capital cost breakdown for a 50-Turbine wind farm.[6,7]

Components	Cost ($/kW)	Percentage of total cost
Rotor assembly	185	18.5
Tower	145	14.5
Generator	50	5.0
Electrical item, electronics, control, and instrumentation	155	15.5
Transmission, drive train, and nacelle	215	21.5
Balance of station	250	25.0
Total installation cost	1,000	100.0

per megawatt capacity in the USA were $1.5–$2 million for wind compared with $800,000 for natural gas.

10.1.3 Operation and maintenance (O and M) costs

There is a high degree of uncertainty with regard to calculation of the operation and maintenance costs of wind farms. Only a small number of turbines have existed for the full expected lifetime of 20 years. The costs are mainly influenced by turbine age, starting low and increasing over time. It is difficult to obtain a lot of detailed costing because of commercial secrecy on the part of the operators. For a 150-kW turbine, the annual O and M costs have been estimated at approximately 1–2% per year of capital investment cost during the first two years of turbine life, rising to 7% per year for years 16–20 of turbine life.[1]

For a 600-kW turbine, the corresponding O and M costs are approximately 1% per year during the first two year of turbine life, raising to 4.5% per year an advantage due to the economy of scale. The Danish Wind Manufacturers Association compiled data on over 4,400 Danish machines installed in Denmark. The predictions, summarised in Table 10.4, demonstrate the economy of scale due to turbine size and illustrate the significant increase of O and M costs due to turbine aging.[7,8] Other surveys have resulted in corresponding conclusions. Maintenance costs in the range $0.01/kWh to $0.06/kWh relating to energy output have been quoted.[7]

Operation and maintenance costs tend to be labour intensive. A 1994 study gave the figure 44% labour cost for the total O and M budget.[9]

10.1.4 Turbine lifetime and depreciation rate

Various surveys agree that the cost of the turbine/generator units contributes 65%–75% towards the total capital cost of installing a wind farm. Providing the appropriate finance, and the terms under which it is provided, depends on the accurate assessment of the plant lifetime.

In Europe, an economic lifetime of 20 years is often used for wind-energy systems.[4] This follows recommendations from the Danish Wind Turbine Manufacturers Association that 20 years is a reasonable compromise figure for the design of components for wind turbines. More recent studies in the USA have suggested that an operating lifetime of 30 years can be used for systems that have been maintained annually and subjected to 10 year major overhauls of the key components.[10]

The depreciation of value of a wind turbine, as with any mechanical device, is not uniform with time. This is reflected in the projected operation and maintenance costs discussed in the preceding Sec. 10.1.3. For the later years of a machine with a projected lifetime of 20 years, the O and M costs rise roughly fourfold over the early year costs.

As with any large engineering project, it is necessary to insure all the key equipment against failure, damage, and late delivery or any major occurrence that might delay the project. A rough estimate of insurance costs is 1% of the total cost

10.1.5 Cost associated with the financing of wind farm building and operation

The design, building, and commissioning of a wind-energy project, including a wind farm, are capital intensive. Moreover, most of the money has to be provided "up front". Once the plant is in operation, the running costs are mainly labour charges and items such as land rental and machinery maintenance that are regular and predictable. Wind-energy projects are often developed by and integrated into existing fossil fuel generation systems by electrical utilities. A traditional utility incorporating its own wind generation would save fuel or reduce the need to purchase electricity from other generators, as well as providing capacity credit value, described in Sec. 8.2.2 of Chap. 8.

All over the world, governments are encouraging their people to save fossil energy by using renewable energy. In order to encourage the use of renewables, including wind energy, there are many schemes involving taxation advantages, various forms of subsidies, and fixed-term and fixed-price contracts for electricity. The securing of an advance sale contract can be decisive in obtaining money from the finance markets.

In the UK, the use of renewable forms of energy is encouraged and subsidised by a scheme known as the Non-Fossil Fuel Obligation (NFFO). This refers to a collection of orders requiring the electricity distribution network operators in England and Wales to purchase electricity from the nuclear power and renewable energy sectors. Similar mechanisms operate in Scotland, being the Scottish Renewable Obligation (SRO), and in Northern Ireland, known as the Northern Ireland Non-Fossil Obligation (NINFFO). The plan is for the last of the customer orders, placed in 1998, to continue until 2018.[11]

The NFFO was put in place by the electricity Act 1989 under which electricity generation in the UK was privatised. The original intention was to provide financial support to the state-owned nuclear power generation industry but this was enlarged to include the renewable energy sector.

Funding for the NFFO was originally obtained from a levy placed on all electricity consumption in the UK. Since the introduction of the new electricity trading arrangements in 2001, electricity suppliers bid for electricity in competitive auctions with any shortfall in price being funded by the levy. The NFFO has recently been operating a trading surplus, expected to reach £500 m by 2008.[11]

When a wind-energy project is financed by private subscription, the economics depend on the financial perspective taken. How quickly do investors want their investments to be repaid and at what rate of return? A short repayment period and a high rate of return pushes up the necessary price charged for the electricity.

Public authorities and energy planners require that invested capital should be paid off during the expected lifetime of the turbine; that is usually 20 years. By comparison, private investors would wish to recover the cost of the turbine during the length of the bank loan. The interest rates used by public authorities and energy planners would typically be lower than those used by private investors.[2]

The fall in the cost of UK wind energy during the 1990s due to the introduction of the non-fossil fuel obligation contracts is illustrated in Fig. 10.1.[2] In the USA, a financial incentive in developing wind-energy facilities is the Federal Wind Production Tax Credit (PTC) introduced in 1994. This provides a substantial economic benefit that lowers the cost of wind-generated electricity for the customer. The cost of financing is also falling as lenders gain confidence in the technology and in the market results.

10.1.6 *Wind regime at the turbine site*

The main consideration in the location of a turbine or a wind farm is the velocity and the duration of the wind regime. Nothing can substitute for a deficiency of wind. Wind power, which is mainly dependent on wind speed, is plotted against wind duration in Fig. 4.2 of Chap. 4. This is the most important form of characteristic relation in all wind-energy considerations. Every turbine site has its own unique version that is characteristic of the individual site. The effect of mean wind speed on the generation cost per unit (kWh) of electricity produced is shown in Fig. 10.2.[2] At a typical wind speed of 8.5 m/s (8.5/0.447 = 10 mph), the UK cost of producing wind-generated electrical energy in 2007 was about 4.5 p/kWh if the capital outlay is calculated to be recovered at 8%. If a capital return is required at 10%, then the generation cost rises to about 5 p/kWh. It is obviously best to operate at the highest possible mean wind speed and to consider

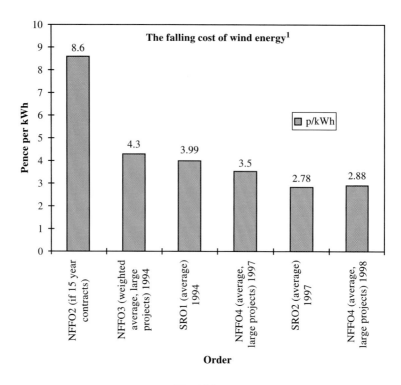

Fig. 10.1

investment in any form of technology that will enhance the wind speed at the turbine.

All of the studies investigating the effect of mean wind speed on the economics of generation give results of the form shown in Fig. 10.2. A typical example from the European Wind Energy Association (not shown here) is to be found in Ref. [3]. Values in $/kWh or £/kWh or €/kWh vary in scale between installation but the form and the trend are common with Fig. 10.2.

10.2 Comparative Costs of Generating Electricity from Different Fuel Sources

It is difficult to produce costings for the generation of electricity from different fuel sources. The generation costs are often produced from sources that have a vested interest in promoting some particular fuel source. Great

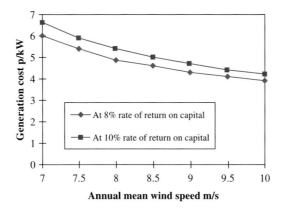

Fig. 10.2   Generation cost of electrical energy versus wind speed.[1]

difficulty arises in attempting to compare "like with like". The fossil fuels are known to have environmental side effects, notably the production of greenhouse gases, that are believed to contribute to global warming. Wind-generated electricity contributes only slightly to air pollution and to gaseous emissions. There have been various attempts to assign numerical values or coefficients to environmental aspects of electricity generation but these are speculative and controversial.

A comparison in qualitative terms is given in Fig. 10.3, which indicates the environmentally friendly nature of wind-generated electricity.[12] In some references, the features such as the environmental impacts are defined as "external costs" (or savings). They may have significant costs or cost avoidances in the overall picture but are not quantified in numerical terms. Few mechanisms currently exist to internalise these costs and the total cost is highly uncertain.

In addition to the pollution and greenhouse gas features of Fig. 10.3, there are other possible external costs such as military expenditures to ensure access to fossil fuels, remediation of polluted sites, destruction of wildlife habitat, and loss of scenery/tourism.[13] None of these items can be quantified in numerical terms and they are all subject to political decision making. Further difficulties with regard to the comparative costs of generating electricity are due to uncertainties in the supply and cost of fossil fuels, and especially oil. In Britain, the price of oil has doubled in the past three years (2010).

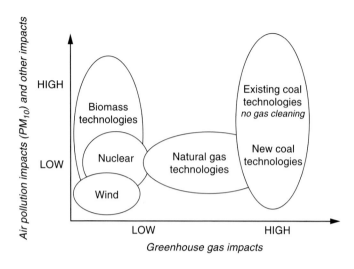

Fig. 10.3 Environmental impacts of generating electricity from various fuels.[12]

In the UK, the NFFO, discussed in Sec. 10.1.5, represents an attempt to incorporate some of the hidden costing/savings of wind generation and to compare the different fuel sources on a "level playing field". Some comparative data for wind generation in comparison with coal and nuclear is shown in Fig. 10.4.[5] Onshore wind generation cost is seen to be of the same order as that of cleaned coal and significantly cheaper than nuclear power. Other prices comparisons, both of that period and more recently, would dispute the conclusion.

An inventory of the wind-power developments in Europe and many countries outside Europe, including the USA and Canada, indicates a continued growth in installed wind-generation capacity. At the end of 2006, the contribution to the world's electricity consumption by wind energy was 0.8%. This is expected to rise to 4% in 2015 because the annual growth of wind-generated electricity, now 30%, is much higher than the global increase in electricity demand (3–3.5%).

At the end of 2005, 2.8% of European electricity demand was supplied by wind generation. The European Wind Energy Association (EWEA) suggests a scenario that the wind contribution will raise to 5.5% in 2010, 13% in 2020, and 23% in 2030.[14] All previous scenarios of the EWEA have been exceeded in performance so that the above figures may be regarded as conservative.[12]

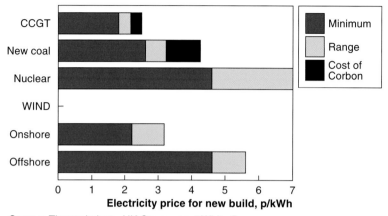

Fig. 10.4   Electricity prices for new build sources of generation.[5]

If the growth of installed wind capacity and electricity consumption are realised at the rate suggested above by EWEA, and corroborated by the European Commission, the expectation must reasonably be continued cost reduction.

A parameter known as the progress ratio (PR) expresses the rate at which unit costs decline each time the cumulative market size doubles. For example, a PR of 0.8 implies that after one cumulative doubling of installed capacity the unit costs are only 80% of the original costs.

For the onshore wind-power development to date, the PR has been evaluated to be 0.82%. The cumulative development of installed wind-power capacity is shown in Fig. 10.5 and seen to double every three years. Based on a less optimistic value PR = 0.9, it is expected that new onshore wind-energy plants in coastal regions will break even with coal-generated power by 2010.[10]

Installation costs for wind power are now (2010) around US$1.8 million/MW for onshore development and between $2.4 million–$3 million/MW for offshore projects. This translates to $0.05–$0.09 per kWh making wind competitive with coal at the lower end of the range. With subsidies, as enjoyed in many countries, wind power becomes cheaper than coal.[1f]

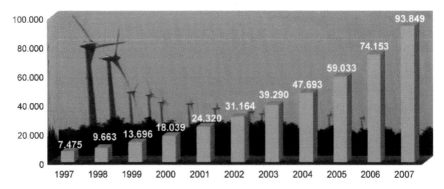

Fig. 10.5 World wind energy installed capacity (MW) 1997–2007 [quoted from the World Wind Energy Assoc. (WWEA)].

Is it possible to suggest that renewable energy, in the form of wind energy, could become economic within about 10 years from now—say by 2020? This suggestion is likely to be very unpopular both with the renewables industry, who would lose their subsidies and with the nuclear, the natural gas, and the coal industries, in the UK. Considerable savings of scale can be expected from the worldwide renewable industry, as it continues to expand. A 10% cost reduction can be expected from any type of equipment if the manufacturing capacity is doubled. It is therefore reasonable to suppose that wind turbine prices will continue to reduce. The wind source remains free. The combination of reducing turbine costs and escalating fossil fuel prices presents the real possibility that wind energy might become the energy source of choice.[16]

References

1. Renewable Sources of Energy with Special Emphasis on Wind Energy, United Nations Department of Economic and Social Affairs, available at http://uneprisoc.org/Wind Energy/UNreportwind.pdf, February 1998.
2. BWEA, *The Economy of Wind Energy*, British Wind Energy Association (BWEA), London, England, 2007.
3. EWEA, *Economics of Wind Energy*, European Wind Energy Association (EWEA), available at http://www.ewea.org/index.php?id=201, 2008.
4. WEC, Renewable Energy Resources: Opportunities and Constraints 1990–2020, World Energy Council (WEC), London, England, 1993.

5. BWEA, *The Economics of Wind Energy*, Information Fact Sheet, British Wind Energy Association (BWEA), London, England, 1997.
6. Renewable Energy Annual 1996, DOE/IEA Report 0603 (96), Washington D.C., the USA, April 1997.
7. Manwell, J. F., J. G. McGowan, A. L. Rogers, *Wind Energy Explained*, John Wiley & Sons Ltd., Chichester, England, 2002.
8. Lemming, J., P. E. Morthorst, L. H. Hansen, P. Andersen, P. H. Jensen, O and M Costs and Economical Lifetime of Wind Turbine, *Proceedsings European Wind Energy Conference*, pp. 387–390, 1999.
9. Spera, D. (Ed.), *Wind Turbine Technology*, ASME, New York, the USA, 1994.
10. Renewable Energy Technology Characterisations, US Department of Energy/Electric Power Research Institute (EPRI), EPRI Report: TR—109496, Washington D.C., the USA, 1997.
11. Wikipedia, Non-Fossil Fuels Obligation, available at http://en.wikipedia.org/wiki/NFFO, July 2008.
12. Van Kuik, G., B. Ummels, R. Hendriks, Perspectives on Wind Energy, *Conference on Sustainable Energy Technologies*, Dubrovnik, Croatia, pp. 75–98, September 2006.
13. Wikipedia, Wind Power, March 2008.
14. EWEA, Nofuel briefing, Feb. 2006.
15. Nature, Electricity Without Carbon, (454), pp. 816–823, Aug. 2008.
16. Swift-Hook, D., 'Reason to Believe', Engineering and Technology, IEE, London, UK, July-August 2008.

Answers to the End of Chapter Problems

Chapter 2

2.1 Differentiate (2.20) w.r.t. (V_2/V_1) and equate the derivative to zero.
2.2 Sine $P \alpha V^3$ doubling causes a $2^3 = 8$ times increase of P.
2.3 In Fig. 2.9, the vertical projection is about 40% of the distance from 10^4 to 10^5.
Estimate A = 30,000 m², D = 195 m.
2.4 See Sec. 2.4.4
r = 90 ft = 27.43 m
V = 20 mph = 8.94 m/s
ω = 1.96 rad/s = 18.7 rpm
2.5 TSR = 7.61
2.6 (a) 0.26 – 0.45 per unit
(b) 0.084 – 0.26 per unit
2.7 If $\eta = 0.25$, D = 4.45 m.
2.8 If $\eta_g = 0.75$, $C_p = 0.35$ and there is no gearbox, D = 12.12 m.
2.9 D = 3.65 m (12ft), when $\eta = 0.25$.
2.10 V = 8.61 m/s = 19.3 mph
2.11 Let the overall efficiency be 25%.
(a) V = 6.26 m/s = 14 mph
(b) TSR = 1.34
(c) D = 3.54 cm = 1.39 in
2.12 (a) T = 3.54 × 10⁶ Nm
D = 68.4 cm = 26.93 in
(b) See Sec. 10.5.2

(c) TSR = 6.46
   $D = 84.2$ m (276.3 ft)
(d) $\eta_g = 0.95$, $\eta_{gb} = 0.9$, $C_p = 0.351$

2.13 $d = 12$ in $= 0.304$ m
   $T = 0.305 \times 10^6$ Nm

2.14 $N_{max} = 35$ rpm
   Propeller is feathered (turned into the wind) to limit rotational speed. Excessive speed would cause large centrifugal forces on the blades plus possible bearing damage.

2.15 (i) Air speed not affected.
   (ii) Ground speed is increased (America to Europe)
      Ground speed is decreased (Europe to America), by 100 mph.

2.16 Air speed = 445 mph
   Ground speed = 445 − 95 = 350 mph
   Time = 3400/350 = 9.71 hours

Chapter 4

4.1 See Table 4.4
4.2 See Fig. 4.5 and Fig. 4.8
4.3 The middle section of the country, to the west of the great lakes. The windiest states are N. Dakota and S. Dakota. See Fig. 10.7,
4.4 The US Met. Office figures for Cleveland, Ohio show that this city has a mean annual windspeed of 10.9 mph (compared with 10.4 mph for Chicago), Table 4.3,
   — Cleveland is about 10% windier than Chicago. In particular, Cleveland is windier in all the months from October through to Aprill.
4.5 Advantages and disadvantages of wind-powered electricity generation.

Advantages	Disadvantages
prime fuel is free	risk of blade failure (total destruction of the installation)
infinitely renewable	suitable small generators not reaily available
non-polluting	unsuitable for urban areas

in UK the seasonal variation matches electricity demand	cost of storage battery or mains converter system
big generation can be located on remote sites, including offshore	acoustic noise of gearbox and rotor blades
saves conventional fuels	construction costs of the supporting tower and access roads
saves the building of (otherwise necessary)	electromagnetic interference if metal roter used
diversity in the methods of electricity generation	environmental objections

4.6 The USA is the world's number one country in political, financial and military terms. It is also the biggest per capita consumer of energy and a massive importer of Middle Eastern oil. Since the Gulf War of 1992, the USA has become the military protector of Saudi Arabia, the world's biggest repository of oil. Americans have a tradition of cheap gasoline and wish to maintain it. The security of oil supplies and the price to consumers is a dominant issue in US domestic politics. The economic feasibility of all other forms of energy has to be contrasted with the supply and price of oil in the USA

4.7 The Weibull distribution is a statistical probability histogram indicating the probability of wind speeds within a certain range at a particular location. In mathematical terms it is defined by equation (4.1).

4.8 In the Weibull equation (4.1) the terms k and C are described and defined in section 4.2.

4.9 $h(V) = \frac{2}{C^2} V \varepsilon^{-\left(\frac{V}{C}\right)^2}$

4.10 For 4 × 4 arragement the contour is 6000 × 6000 square feet or 3.34 × 10^6 m². This represents an area of 334 hectares or 826 acres.

Chapter 7

7.1 See section 7.2, especially the analysis of equation (7.7) and 7.8)
7.2 If the rms line-line voltage is 415 v the peak phase voltage $E_m = \frac{415}{\sqrt{3}}\sqrt{2} = 339$v.

240 Electricity Generation Using Wind Power

The average value E_{av_o}, from (7.9) is $1.654\ E_m = 560$. From (7.8)

α_e	$\cos \alpha_e$	E_{av}
0	1	560
30	0.866	485
60	0.5	280
90	0	0

7.3 See section 7.3.5

The distortion factor has the value $3/\pi = 0.95$, determined by the ideal waveform of Fig. 7.8(a), f_a for all values of firing-angle α_I.

7.4 Perform a Fourier analysis of the current waveform of Fig. 7.5

7.5 The current waveform of Fig. 7.5 and Fig. 7.8 is defined by the equation

$$i_a(wt) = \frac{E_{avo}}{R} \cos \alpha \Big|_{\alpha+30°}^{\alpha+150°} - \frac{E_{avo}}{R} \cos \alpha \Big|_{\alpha+210°}^{\alpha+330°}$$

Since the rms values of the positive and negative parts of the wave are equal the rms supply current is given by

$$I_a = \sqrt{\frac{1}{\pi} \int_{\alpha+30°}^{\alpha+150°} \left(\frac{E_{avo}}{R} \cos \alpha\right)^2 d\omega t}$$

$$= \frac{E_{avo}}{R} \cos \alpha \sqrt{\frac{1}{\pi}[\omega t]_{\alpha+30°}^{\alpha+150°}}$$

$$= \sqrt{\frac{2}{3}} \frac{E_{avo}}{R} \cos \alpha$$

$$= \frac{3\sqrt{2} E_m \cos \alpha}{\pi R} \quad Q.E.D$$

7.6 587v., 3.24 A
7.7 0, 12051 w, 11840 w, 10133w
7.8 622 v, 37.24 A
7.9 154°, $P_{165} = 0$, 52.6 kVA
7.10 933v, 32.5 A
7.11 146.4°
7.12 0.675

The action of the capacitor is discussed in section 7.3.3.

Index

American Farm Windmill, 42
apparent power, 100–101
Argand diagram, 160
average power, 100–101
axial thrust, 19

back scattering, 216–217
Beaufort scale, 67–69
bending moment, 21–22
Betz' Law, 14
British Wind Energy Association (BWEA), 203

capacity credit, 183, 187–188
carbon dioxide, 209
centrifugal force, 22, 63
Charles F. Brush, 43
Clear Air Acts, 206
complex power, 105–109
complex vector representation, 136–145
Constant Speed Constant Frequency (CSCF), 132–133
Coriolis force, 3
crosswind spacing, 90–91

Darrieus design, 62–63
Dept. of Trade and Industry (DTI), UK, 64
diffuser augmentor, 89
direct current (d.c.) generator, 113–114
direct current (d.c.) link, 149, 151–155
displacement factor, 162–163

distortion factor, 162–163
doubly fed induction generator (DFIG), 134–136, 176–177
downwind spacing, 90–91
Dutch windmill, 4–5

economic aspects of wind power, 222–235
Electrical Research Association (ERA), UK, 59
Electrical Research Development Association (ERDA), USA, 51
electromagnetic interference, 215–217
energy credit, 183–187
Enfield-Andreau turbine, 47–48
European Wind Energy Association (EWEA), 230–233

Federal Wind Production Tax Credit (PTC), USA, 229
fixed speed, directly coupled induction generator, 175–176
forward scattering, 216–217

gearbox efficiency, 29–31, 33
Gedser turbine, 48–49
gyroscopic force, 20–21

harmonic distortion, 115–119
histogram, 79, 82
hub height, 6
Hutter turbine, 49–50

induction generator
　see three-phase induction generator
induction machine, 121–132
induction motor
　see three-phase induction motor
instantaneous power, 99–100
integration of wind farm, 181–202
isobar, 1
isovent, 74
inverter, 154–159
inverter power factor, 162–163

Jacobs Wind Electric, 45

kinetic energy, 9–13
Kirchhoff's Law, 97

lamp flicker, 175
Leonardo da Vinci, 5
lightning strikes, 194
linear momentum, 13
loss of load expectation (LOLE), 188

matrix converter, 166–169
millibar, 7
MOD turbine, 51–52

nineteenth century windmills, 41–43
nitrogen oxides, 207–208
Non Fossil Fuel Obligation (NFFO), 228, 229, 232
Northern Ireland Non Fossil Fuel Obligation (NINFFO), 228
Nysted wind farm, Denmark, 56–57

OPEC (Organisation of Petroleum Exporting Countries), 51
Orkney wind turbine, 35–36, 56–57

Parris Dunn turbine, 43
particulates, 206–207
Pascal, 1, 7
pitch angle, 27–28, 133
polar moment of inertia, 24
power coefficient, 15–17

power factor, 101–103
Poul La Cour, 44
probability density function, 79

Rayleigh distribution, 84
reactive power, 103–105, 109–111
rms current, 160–162
rotating magnetic field, 121

Savonius rotor, 61–62
Scottish Renewable Obligation (SRO), UK, 228
shadow flicker, 219–220
Smith Putnam turbine, 45–46
solidity factor, 22–23
statistical representation of wind, 79–84
sulphur dioxide, 206–207
switched reluctance generator, 145
switched reluctance motor, 143–145
synchronous capacitor, 93
synchronous generator, 114–121
synchronous motor, 93, 120–121
Systeme Internationale (S.I.) Units, 8

teeter, teetering, 22, 63
three-phase bridge inverter, 154–163
　　output voltage, 154–159
　　average power, 158–159
　　reactive power, 159–160
　　rms current, 160–162
　　power factor, 162–163

three-phase bridge rectifier, 148–154
three-phase cycloconverter, 165–166
three-phase induction generator, 127–132, 199–201
three-phase induction machine, 121–132
three-phase induction motor, 122–127
tip speed, 17, 29
Tip Speed Ratio (TSR), 17–18, 26–28
torsional stress, 24–25, 34–35
turbine lifetime and depreciation, 227
turbine operation and maintenance costs, 206–207
twentieth centry wind turbines, 43–50

US Dept. of Energy (DOE), 57
US Meteorological Office, 76, 92

variable Speed Constant Frequency (VSCF), 133–134
variable speed, direct-drive synchrous generator, 177–178
voltage flicker, 194–195

Weibull function, 82–84, 92
Whitelee wind farm, 55–56, 59, 91
wind turbine financing, 228–229
wind turbine noise, 209–215
    aerodynamic noise, 212–214
    mechanical noise, 214–215

yaw, 20, 24